Synthesis Lectures on Engineers, Technology, & Society

Volume 29

Series Editor

The mission of this Lecture series is to foster an understanding for engineers and scientists on the inclusive nature of their profession. The creation and proliferation of technologies needs to be inclusive as it has effects on all of humankind, regardless of national boundaries, socio-economic status, gender, race and ethnicity, or creed. The Lectures will combine expertise in sociology, political economics, philosophy of science, history, engineering, engineering education, participatory research, development studies, sustainability, psychotherapy, policy studies, and epistemology. The Lectures will be relevant to all engineers practicing in all parts of the world.

Camilo Andrés Navarro Forero

Untangling Complexity—Peace Building Engineering

Systems Thinking & Complexity Management to Support Community Development

Springer

Camilo Andrés Navarro Forero
Fundación Universitaria Salesiana
Bogotá D.C, Colombia

Universidad de los Andes
Bogotá D.C, Colombia

Universidad Sergio Arboleda
Bogotá D.C, Colombia

University of San Diego
San Diego, California, USA

ISSN 1933-3633 ISSN 1933-3641 (electronic)
Synthesis Lectures on Engineers, Technology, & Society
ISBN 978-3-031-37037-3 ISBN 978-3-031-37038-0 (eBook)
https://doi.org/10.1007/978-3-031-37038-0

This Springer imprint is published by the registered company Springer Nature Switzerland AG
The registered company address is: Gewerbestrasse 11, 6330 Cham, Switzerland

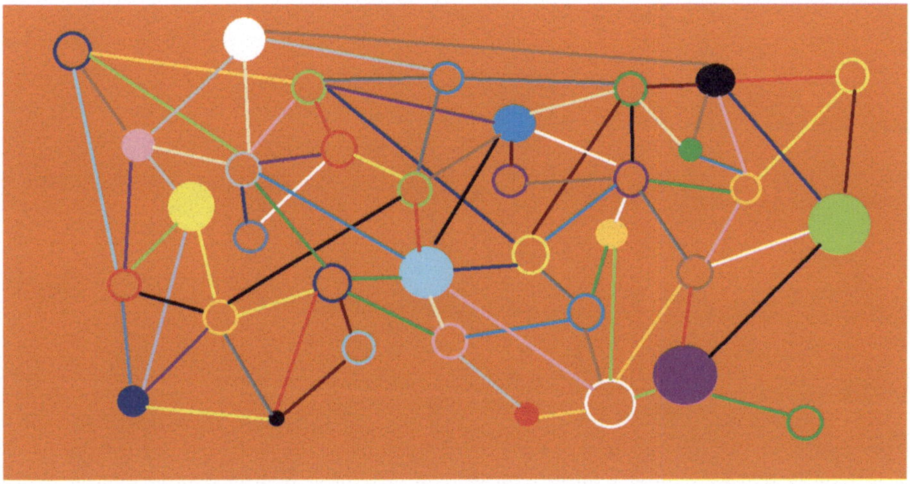

A complex world Camilo Andrés Navarro Forero original drawing 2022

Generating and maintaining Peace, within communities, nations, and the world is a vital component for achieving human fulfillment and happiness.

With gratitude and appreciation for my family, friends, my students, my colleagues, and all those who should profit from what was written in this book.

Foreword 1

I have had the privilege of knowing and working with Camilo Navarro, Ph.D. for the last several years. Camilo is a remarkable, innovative leader. He inspires students, colleagues and the community in his approach to bringing the world of peacebuilding and engineering together to create real impact and change. He is a person who "walks the talk," setting an example for all of us in how his skills, knowledge and passion can truly create a better world.

I first met Camilo when he was an International Visiting Scholar at University of San Diego's Shiley-Marcos School of Engineering. As I was working and teaching at University of San Diego's Kroc School of Peace Studies, Camilo introduced me to a new perspective on how engineering can be such a critical tool for sustainable peace building. I have spent many years teaching social innovation and social entrepreneurship in Colombia, collaborating with several universities there. I see first-hand in Colombia and globally the need and opportunity to connect engineering with social impact and economic empowerment. I continually learn from Camilo. Starting with his work leading Engineers without Borders in Colombia, to inspiring his students to think about problem solving in truly transformative ways, he is a role model. One particular project where Camilo truly "walked the talk" is with his graduate students when he applied his engineering and problem solving skills to help launch and scale a unique brewery in Colombia "Cerveceria la Roja", co-founded with former members of the FARC-EP guerillas. This brewery proved that second chances are real and by bringing passion, collaboration and the right tools and technology to the opportunity, real bridges can be created. Cerveceria la Roja and so many more projects he supports that provides employment for those who are challenged finding jobs, and with their success, they are winning accolades across the region for their fabulous beer products.

As Dean of Engineering at Salesiana University, Camilo Navarro is inspiring and leading a new generation of engineers committed to bringing their skills and knowledge to support change and impact in Colombia and beyond. He is bringing together diverse stakeholders within and beyond the university to engage students and the community in projects and new approaches linking engineering and peacebuilding. Karen Henken lives in San Diego California United States of America.

Karen Henken, M.B.A.
Visiting Professor
University of California San Diego
San Diego, CA, USA

Visiting Professor
CESA (Colegio de Estudios Superiores de Administración)
Bogotá, Colombia

Foreword 2

In this Foreword the importance of the book *Peace Engineering Untangling Complexity Systemic Thinking & Complexity Management to Support Community Development Complexity Funnel Methodology CFM and Social Transformation Workshops STW* will be addressed and the reason why it was written. The idea is to stimulate the readers interest in the book. Caroline Baillie, Ph.D.

Jerry McCann brings his depth of understanding and experience in both engineering and peacebuilding to peace initiatives around the world.

Jerry McCann is currently Senior Advisor for Build Up, a global non-profit working besides local organizations through participatory interventions, research, and training to combine peacebuilding best practices, participatory methodologies, ad digital technologies. Before his 20 years of work overseas in development and peacebuilding, Jerry was a structural engineer and construction manager in California. His unique profile and deep commitment to contextually driven, locally owned processes of social change have allowed him to demonstrate the ways in which engineering, management science, and peace building can not only compliment but elevate peace, development, and stability in societies in or recovering from conflict.

Jerry is now based in San Francisco but continues to work around the world. He is a digital nomad with a keen interest in engaging with individuals and groups that recognize the only solution for peace comes from within supported, when sought, by greater external understanding and equitable partnerships.

Jerry is also involved with academic work with the Engineering and Peace Master's Program at Drexell University and work with the University of San Diego.

Jerry McCann
University of San Diego
San Diego, California, USA

Preface

Engineering, as a concept, has been around since ancient times, however, engineering as we know it today is defined by what's been achieved over the last 150 years. One leading edge of those achievements has been in the military where engineers have played critical roles. Contrarily, engineers have rarely been considered for roles they can play in the pursuit of peace. That is now changing. Recently, the importance of bringing the concepts of peace and engineering together has emerged and with it the new field of *peace engineering*. As with any new field, this field needs to be defined, developed and supported with tools and competencies that will allow engineers to actively engage in and support the initiatives of the peacebuilding community, especially given the negative role engineers can play in isolation.

One of the ways that engineering has evolved in modern times has been to become more and more isolated and protected from the social, political and environmental impacts of engineered solutions. Much of this isolation has come from what is justified by the engineering community as a professional necessity to protect against liabilities in a litigious world; many of these liabilities are outside what engineers believe they should be held accountable for. The engineering profession argues that if engineers are being hired to provide technical solutions, their responsibility is in the rigor and execution of those technical solutions, not in the intended and unintended consequences that follow. This has created dangerous cumulative precedence, especially as the world has become more connected and humanity's role on the planet has become more complex and impactful. The complex social systems that exist in communities, in regions, in nations and in the world must be considered and included in the responsibility of engineers. However, at the same time it is important that engineers do not treat conflict within or between communities, at any scale, as a system that can be solved through engineering processes of modelling and simulation as some in the emerging field of peace engineering believe is possible.

Systems engineering has become an important aspect of engineering that recognizes the importance of more holistic and interdisciplinary solutions to complex problems affecting a system. It is through the evolution of this field of engineering, as well as because of a growing number of engineering academicians and practitioners who feel that not enough is being done to end violent conflicts in the world, that an engineer-led, systems-based approach to peacebuilding has emerged and is being promoted by its advocates as the basis of peace engineering. The danger with this approach is that engineers assume they have the ability to model conflict, and then solve the conflict within the closed system of that model. This approach has been largely rejected by the peacebuilding field, creating competing efforts rather than the needed collaborative, interdisciplinary approaches many in the peacebuilding field advocate for. While those advocating for engineer-led, systems approaches highlight the importance of engaging other disciplines, what is really needed is transdisciplinary approaches, led by those with the deepest understanding of the conflict and the socio-political dynamics shaping it. It is only through a peacebuilding lens, one that by design is an iterative and adaptable learning process that peace can be achieved, consolidated and sustained.

As you will find in the pages that follow this preface, what is needed are opportunities to bring participatory processes, focused on stakeholder engagement, into the concept of systems modelling. This requires systems engineers to be guided by community facilitators practiced in the processes of stakeholder engagement, as well as by socio-political researchers, using methodological approaches such as the creation of safe spaces for dialogue and participatory action research (PAR), to ensure that the approach to modelling the systems that make up a society in conflict is, in itself, a peacebuilding process. In other words, it is not the engineers who should be defining the complex dynamics of communities in conflict but rather those skilled in doing so to ensure the modelling reflects the realities of the broadest spectrum of stakeholders in the communities. Peace engineers who ignore this fundamental necessity will quickly find that their efforts lack legitimacy in the eyes of the stakeholders and will likely be seen as unwanted external interventions. Ultimately, solutions to peace are not formulaic, regardless of how complex the modelling, simulation and problem-solving processes are, and unless those central to consolidated, sustained peace, namely the stakeholders and the broader population, are fully integrated into the process, it is likely to fail.

The theme that runs through this book and through the peace engineering we advocate for is that while we cannot engineer peace, engineers can play an important supportive role in the peacebuilding processes that lead to peace.

Acknowledgments

Fundación Universitaria Salesiana, Bogotá D.C Colombia Latin America
Universidad los Andes, Bogotá D.C Colombia Latin America
Universidad Sergio Arboleda, Bogotá D.C, Colombia Latin America—
University of San Diego California United States of America
Engineers Without Boarders Colombia
Engineering Social Justice and Peace
Engineering Exchange for Social Justice
Humanitarian Engineering Program Universidad Sergio Arboleda—Colorado School of Mines.
María Catalina Ramírez, Advicer, Mentor and Friend
Caroline Baillie
Odesma Dalrymple
Jerry Mccann
Karen Henken
Alex Mejia

About This Book

This book is divided into four chapters:

Chapter 1

The first chapter considers the development of General Systems Theory, Complexity Theory, Systems Thinking, Critical Systems thinking and how all these concepts can be applied to support decision makers in social complex systems (SCS). We must begin by talking about Systems Theory and Complexity Theory because this theoretical background is necessary to understand the language of Systems Thinking and the key concepts it employs. Systems approaches often start with the help of Social Theories which is important to the success or failure of engineering interventions. Systems and methodologies are strongly related to the social theories they ratify. The first chapter explains essential systems theory so that the reader can understand the practical systems methodologies and their applications that come later.

Chapter 2

The second chapter explains the development of an innovative pluralistic multi-methodology called the Complexity Funnel Methodology (CFM) that seeks to help decision makers in social complex systems. This methodology takes the form of a series of Social Transformation Workshops (STWs) that map the current state of the social complex system, with emphasis placed on the identification of actors, the interrelations between the actors and the resulting impact of the relationships on various variables. The goal is then to reduce the complexity within the system measure in terms of obstructions, both discrepancies and conflicts. In Sect. 2.3 we will present the social entropy metric that will allow us to track how well the workshops are reducing complexity within the system. We use different systems methodologies and mixed methods together to confront multiparadigm systems in a theoretically informed and very creative way with the active participation of the community with the final objective of legitimizing interventions and assuring project's continuity over time.

Chapter 3

Explains the STW workshops which are intended to help improve the leadership, organizational and managerial skills of decision makers in social complex systems showing the impact of engineering tools while supporting community development projects.

Social Transformation Workshops
STW#1 "The Web of Life"
STW#2 "Rich Picture"
STW#3 "Critical Systems Heuristic"
STW#4 "Social Cartography"
STW#5 "Viable Systems"
STW#6 "Organizational Pathologies"
STW#7 "Systemic Arquetypes"
STW#8 "Idealized Design"
STW#9 "Conceptual Model" SSM Checland P
STW#10 Gaps and Priorization
STW#11 Design or re-design a Project
STW#12 Matematical and Dynamic modeling.

Chapter 4

Application—Systems Practice

The fourth chapter of this book talks about systems practice and the way that systems ideas can be put to use in dealing with problems posed by complexity that shows the application of different systemic thinking tools with a systems approach for all types of complexity (technical, process, structural, organizational, people and coercive complexity).

We strongly believe that different approaches to deal with complexity will address different aspects of it. Each of those aspects is motivated by different social theories and philosophies and they all have both strengths and weaknesses because of the theoretical assumptions that were made while using them. We also believe that the best way to address complexity in social complex systems is if we use them in combination, compensating for their weaknesses and focusing on their combined strengths. This way of exploring complexity is called Critical Systemic Thinking.

We will share the application of the multimethodology in a five phase program starting with private companies, then public companies while working with communities in a national and international level. We will finish with the application of the methodology in peace building communities in Colombia.

Phase 1 Private Companies
Phase 2 Public Companies
Phase 3 Communities
Phase 4 International Communities
Phase 5 Peace Building Communities.

Introduction

Systems Thinking seems to assume that if you are trying to change a system you must have some type of power or authority within that system. But the reality is that community organizations, rarely have formal power over the thing they are trying to change. The systems approach encourages the employment of a variety of methodologies in combination to manage "messes" and "wicked problems" that may emerge from a collection of interacting objects, both actors and variables. General Systems Theory and particularly Critical Systems Thinking works by mixing methods and methodologies which presents a helpful tool for decision makers in social complex systems. Engineering as a profession can contribute to systems of injustice or be part of a systemic change towards social justice. However, those decision makers developing and enacting engineering projects, processes, and systems, are rarely educated in the complexities of the local and global social/economic and political contexts in which they are working. Given the right education and working together across disciplines and with community grassroots organizations engineers have the potential to provide knowledge and skills where it matters most, to co-create just and ecologically sustainable socio-technical solutions. The system's approach gives all actors the possibility of presenting their perception about a problem making the voices heard and giving important information to those decision makers. In turn, this allows decision makers to achieve successful outcomes by becoming multi-methodological. In this way we say that a systemic thinking approach can help in promoting Social Justice and Peace Building.

In engineering, there is a large variety of systemic tools that aim to provide information on the systems to which they are applied. Many of these tools are used separately and the synergy between them is not used to propose integrative viable solutions. The 21st century comes with a lot of new challenges; and with it a new skill set that is needed to face these new challenges. It has become increasingly evident that the problems and challenges we face these days are highly interlinked, complex and require a pluralistic multidisciplinary approach. This book brings such thinking to the forefront and focuses the discussion on the development of a new multi-methodological proposal. This new multi-methodology proposal wants to help decision makers in complex social systems who have the difficult task of making decisions that affect many people based on incomplete information and limited resources to optimize the use of those resources giving continuity and legitimacy to projects over time. Thus offering relevant tools when the complexity of the situation makes it difficult for decision makers to really know what to do and how to do it.

Contents

About the Author

Camilo Andrés Navarro Forero is a Colombian Ph.D. in Engineering, he is Dean of Engineering at Salesiana Catholic University in Bogotá Colombia where he oversees the Agro-industrial, Sustainable Energies and Informatic Engineering Programs, and the Data Management Technology program. He was Chair of Organizational Engineering at Sergio Arboleda University in Bogotá Colombia, where he was a tenured professor for five years. He is an industrial Engineer from the Universidad de los Andes in Colombia, MS Industrial Engineering Magister with an emphasis in Organizational Management and MSS Peace Building Magister and Ph.D. in Engineering from the same university. He works as part of the Engineering Exchange for Social Justice in the University of San Diego California USA were, he supports the User Centered design class in the Shiley Marcos School of Engineering. He is an active member of Engineers without borders Colombia, The international group of Engineering Social Justice and Peace and The group of Humanitarian Engineering from the Universidad Sergio Arboleda in collaboration with the Colorado School of Mines. He has national and international experience working with vulnerable communities and most particularly with peace building communities in Colombia where he studied the success of engineering interventions to support projects within these communities. He teaches Introduction to Engineering, General Systems Theory, Systemic thinking, Systems Dynamics, Human Factor Engineering and Project management courses.

List of Contributors

Students

Gabriel Paris
Emiliano Aristizábal
Daniela Rincón
Diego Pacheco
Sofia Castillo
Mauricio Zabala
Martin Prieto
Valentina Perdomo
Natalia Martinez
Camila Espinoza
Ximena Escobar
Daniel Sierra
Katherine Schifino

Professors

Maria Catalina Ramirez
Caroline Baillie
Odesma Dalrymple
Alex Mejia

Abbreviations

CFM	Complexity Funnel Methodology
CSH	Critical Systems Heuristic
ID	Idealize Design
IP	Interactive Planning
MM	Multi-methodologies
OR	Operational Research
SCS	Social Complex Systems
SSM	Soft Systems Methodology
STW	Social Transformation Workshops

List of Figures

List of Photos

List of Tables

1 Important Definitions

1.1 Peace

Recent academic research on Peace has overwhelmingly focused on negative peace, or the absence of conflict, aggression, violence, and war as described by Galtung (1996).

Given the wide range of definitions and approaches of the concept of "*Peace*" that currently exist, and since we do not want to enter into a philosophical argument, we will use the concept of positive peace developed by Galtung (1996) and more deeply by Webel and Galtung (2007a, b). Both authors define Positive Peace as:

> "*More than the absence of violence, it is the presence of social justice through equal opportunities, a fair distribution of power and resources, equal protection, and the impartial application of the law.*"

Likewise, Vesilind argues that positive peace.

> "*...would establish social equity and justice, economic equity, and ecological balance, protecting citizens from attack and meeting basic human needs.*"
>
> (Vesilind 2010).

1.2 Peace Engineering

"*Engineering Peace Peacefully*" means making "*Peace by peaceful means*" as mentioned by Galtung (1996), it means that as engineers we are always careful on the way our engineering interventions affect a community and we want to be sure that the intervention

C. A. Forero Navarro, *Untangling Complexity—Peace Building Engineering*,
Synthesis Lectures on Engineers, Technology, & Society 29,
https://doi.org/10.1007/978-3-031-37038-0_1

always improves the community well-being and never affect them in a negative way. Thinking carefully about how their designs could help to foster a lasting peace. Making social cohesion profitable and sustainable. Connecting the dots between peacebuilding and engineering, investing in positive social change.

It is not enough to support Peace Building Communities with good intentions, been a good Peace Engineers requires that engineers are always aware that you are doing it on a peaceful way.

1.3 General Systems Theory

There is not a general agreement on '**Systems Theory**'. There are different notions of defining systems and a variety of systems principles. Although the system concept is very widespread, they are multiple definitions that co-exist.

1.3.1 System

'*A set of objects together with relationships between the objects and between their attributes*'

(Hall and Fagen 1956).

From this definition follows that it is difficult to point out something in the social world that is not a system.

There is an absence of a well-articulated theory relevant to all systems problem related situations. But there are a lot of different frameworks and theories and related methodologies.

Donella Meadows (2004).

1.3.2 Systems Science

Systems science is the ordered arrangement of knowledge acquired from the study of systems in the observable world, together with the application of this knowledge to the design of man-made systems (M'Pherson 1974).

Systems science does not aim to find the one true representation for a given type of systems, but to formulate general principles about how different representations of different systems can be constructed to be effective in problem-solving (Heylighen 1990).

Systems science is a science whose domain of inquiry consists of those properties of systems and associated problems that emanate from the general notion of system hood (Klir 2001).

Systems science can be defined as the scientific exploration and theory of systems in the various sciences, such as biology, sociology, economics, etc., while general system theory concerns the principles that apply to all (Strijbos 2010, p. 454).

1.3.3 Social Systems

Ackoff and Gharajedajhi definition for social systems: (Ackoff and Gharajedaghi 1996).

"*Systems that have purposes of their own, contain parts (other social systems) that have purpose of their own, and are usually parts of larger social systems that contain other social systems*".

Doffer definition for **Social System**: "A group of people who relate to each other, according to certain legal rules of organization and who shared the same culture or civilization in a specific space or time" (Ackoff 1999a, b).

1.3.4 Simple System: One Cause One Effect

Normally single isolated, well-defined cause led to a single isolated well-defined effect, this is as simple as it gets, one cause one effect. Playing rugby, driving a car, and even simply taking a walk are all everyday examples of Newton's laws of motion. Compiled in 1687 by English mathematician Isaac Newton, the three main laws describe forces and motion for objects on Earth and throughout the universe. For more than 300 years thanks to the work of People like Isaac Newton we are able to accurately predict and confirm by simple measurement the speed and direction of an object depending on the forces that act on it. One cause one effect.

Other example: The *Staphylococcus aureus bacteria*, which normally rests without any other consequences on the surface of your skin, is pushed deep into you when the stinger of that buzzing bee stung into your hand. The redness in your skin is spreading, slowly, up your arm. No problem, the antibiotic dicloxacillin that your doctor prescribes kills the bacteria, and over the next few days the redness fades away. The cause of the infection was clear, the cause of its successful end was equally clear.

If we pour an equal amount of the solutions of an acid and a base of equal strength together and so they neutralize each other. The solution pH is 7.0, exactly, you measured it with your digital pH meter. Just what you learned about in chemistry, just what you expected.

One cause one effect it looks easy.

Unfortunately, the world we live in I not as simple as the replication of this experiments, we cannot just simply measure to predict what is going to happen.

1.3.5 Complex Systems and Emergent Systems Properties

> Complex systems can be characterized as having a large number of components or subsystems that participate in many loosely structured interactions the outcome of which is not predetermined
>
> "Such systems adapt and evolve over time as they are affected by their own purposeful parts and by the turbulent environments in which they exist."
>
> (Jackson and Keys 1984).

In Simple Systems, a clearly identifiable single cause, leads to a clearly predictable and identifiable single outcome it is easy it is simple. This does that and we can replicate and predict. It works every time, you could even forget about the rest of the universe. Simple. We are so used to seeing, or thinking that we are seeing, such simple examples, that we can come to think that this is the way the entire world works. A rational and universal reductionist approach to everything, but t is not. The world we live in is more complicated than ever and the problems we are facing are more complex than ever before a simply traditional rational linear reductionist approach will not be enough.

Real physical, biological, and social entities are quite different from those "*simple*" examples mentioned before. They are true complex systems, meaning that they consist of a large number of strongly interacting elements.

> "*... The properties of the whole system "emerge" or "self-organize" from its parts and yet those macro system properties are typically strikingly different from the properties of the micro interactions between its parts.*"
>
> (Kelso 1995).

Positive and negative "*feedback loops*" of interactions can make it hard to untie the tangled ball of effects and so we may not be able to identify a single cause with a single effect. This has been well documented in the politics of international relationships (Goldstein and Freeman 1990; Jervis 1997).

The cause-and-effect relations are even more difficult to discern if there are also subsystems interacting within a larger system, with contrasting types of interactions or goals.

Such systems may have "**Attractors**," a limited number of sets of values of the variables that the system returns to, if disturbed. This concept can be related to resilience.

1.3.6 Complex Adaptative Systems

These systems are formed by agents that are capable of learning and adapting as they interact with other agents. In complex social adaptative systems, the interactions of the agents generate emergent and self-organizing behavior.

For Mitchell, a complex adaptative system is "*a system in which large networks of components with no central control and simple rules of operation give rise to complex collective behavior, sophisticated information processing, and adaptation via learning evolution*". (Mitchell 2009).

Complex adaptive systems (CAS) are dynamic and self-organizing systems composed of numerous interconnected elements or agents that interact with each other and their environment. These systems exhibit emergent behavior, meaning that the collective behavior of the system arises from the interactions and adaptations of its individual components.

1.3.7 Adaptation

Adaptation is the capacity for a system to change its state in response to some change within the environment.

The system does this in order to optimize its state within that environment according to some metric (Navarro 2001).

The process of adaptation refers to the adjustment, modification, or change in response to new or changing circumstances or conditions. It is a fundamental characteristic of living organisms and complex systems that enables them to survive, thrive, and maintain stability in dynamic environments.

1.3.8 Self Organization

If a system shows organization, we tend to assume that someone or something must have design it in that order. Self-Organization is the idea that global coordination can be the product of local interactions.

A self-organizing system is a complex system that has the ability to spontaneously generate patterns, structures, or order without external control or central coordination. It is a process by which global organization emerges from the interactions and behaviors of the system's individual components. In a self-organizing system, the interactions between the elements or agents within the system give rise to collective behavior and patterns that are not explicitly directed or imposed by an external entity. The system's organization emerges from local interactions and feedback loops among its components.

1.3.9 Social Complex Systems (SCS)

"Social complex system refers to a system that involves interconnected individuals or entities and exhibits complex behavior arising from their interactions. These systems can encompass social, economic, political, or cultural aspects of human society. Social complex systems are characterized by emergent properties, non-linear dynamics, feedback loops, self-organization, and adaptation. They involve the interplay of various factors such as individual behaviors, social norms, information flows, power dynamics, and external influences. Understanding social complex systems often requires interdisciplinary approaches and analytical tools from fields such as sociology, economics, psychology, network science, and complexity theory.

Social Complex Systems that have a large number of components, often called agents that interact and adapt or learn" (Holland 2006).

> "*Systems that have a large number of elements both actors and variables, interrelated according to certain legal and social rules, culture or civilization in a specific space or time. They interact, adapt, and are affected by their own purposeful parts and by the environment & they learn from those interactions.*"
>
> (Navarro 2020).

Since the eighteenth-century authors like Marx called for action and not just interpretations of this complex systems.

The philosophers have only interpreted the world, in various ways; the point is to change it.

(Marx 1975).

Gibbons et al. (1994) states that they are two types of research relate to systems.

The first type are problems defined by the scientific interest of an academic community and the primary aim is to gain new knowledge for its own sake. It's in Gibbons words … *"a complex of ideas, methods, values and norms that has grown up to control the diffusion of the Newtonian model of science to more and more fields of enquiry and ensure its compliance with what is considered sound scientific practice."*

Tranfield and Starkey (1998) The second type is more focused on the application and its produced to satisfy the demand of particular users it embrace trans disciplinary and aspects of action research because the real world is not divided into academic disciplines and because it is unable to rely on any single discipline to provide to provide the theory to guide an intervention.

Systems practice is therefore of necessity transdisciplinary drawing as seems appropriate on systems concept from different areas as social sciences, philosophy among many others and upon the framework of different systems ideas. And this second type of research is more accurate when you want to support a project as a manager using Systemic thinking tools.

Since the 1940s and 1950s the application of systemic thinking in management has been analyzed primarily in Operational Research OR focusing on clients and their needs and requirements. But more recently problem owners and actors that are affected but not involved. When we operate in type 1 research, we need that the studied complex systems remain stable enough during some period so that accurate measurements can been made. Outcomes and they are detached and can be used in the service of other systems approaches with varying degrees of success or failure, they are different models, tools, and techniques usually associated with the approach.

To compensate the lack of a well-articulated common theory they are several systems methodologies that have emerged, and they carry some theoretical assumptions but are more focused on providing some principles to be able to use methods and techniques to ensure they are achieving their objectives of providing desirable improvements and solutions for the unique and specific circumstances of each intervention. These methodologies that use methods and techniques they transfer problem solving capabilities and they are more applicable than theories and allow us to learn from unique problems situations and then transport the learnings and feedback from past experiences to another problem. As systems practitioners we must embrace action research and we must accept that when you do research on social systems this will inevitably lead to change. It's important that the practitioners take responsibility of the results and outcomes of the implementations of their methodologies and seek to assure that as possible this represents improvements to the current situation. Even if is not a success case practitioners must reflect on the results of their implementations and learn from them so they can improve their methodologies in a continuous improvement cycle for the next time they are used. A good example of this is the Soft Systems Methodology of Peter Checkland who was able to build an exemplary action research program and was successful on continually improve his methodology and then inspire generation of academics.

It's important to talk about the difference between Methodology and Method.

A Methodology refers to the logical principles that govern the use of methods in order that the philosophy or theory embraced by the approach is properly respected and appropriately put into practice in other words it translates the philosophy or theory into practical application.

Methods are concerned with achieving more specific procedural.

1.3.10 Peace-Building Systems (PBS)

We call Peace-Building Systems, complex social systems where Peace-building communities actively participate, among which we can find civilian victims of the conflict, military victims of the conflict, military, demobilized, people in the process of reincorporation as defined by the United Nations, and other actors directly or indirectly connected to this peace building processes.

A peacebuilding system refers to a comprehensive and coordinated approach to promoting sustainable peace in societies affected by conflict or violence. It involves a range of actors, institutions, and processes working together to address the root causes of conflict, prevent the recurrence of violence, and foster long-term peace and stability.

1.3.11 Dialogue

Dialogue is a flexible and adaptable process. It is particularly useful when stakeholders are not ready to sit down and formally negotiate. When violence, hatred, or mistrust is stronger than the will to create a consensus or when there is a clear imbalance of power or a lack of will of the participants, it may be that the situation is not conducive to generating a dialogue. For this space to be successful, participants must feel free to express their thoughts without fear of retaliation or rejection. Dialogue can be the beginning or a complement to more formal conversations.

"*Formal channels are not always useful, and you know it*" Jimmy Carter In the dialogue there are no winners or losers. A negotiation seeks to reach a concrete agreement, the dialogue seeks to create bridges between the different actors that allow sharing perspectives and discovering new ideas.

"*Conflicts are resolved through changes that at first seem unimaginable*" Nelson Mandela It is also important to mention that crises often require an immediate response, and the population may feel frustrated by the delay and by the interminable conversations that do not seem to produce concrete results or actions. However, the global trend is towards increased dialogue and not less. "*... to capitalize on respect for dialogue by working to ensure that mediators and others involved are as well-equipped as possible to generate effective and lasting results*" (HD center 2007).

Another complementary element to positive peace is the agreement, as a collective process where different actors participate to find an answer. This is based on the concept of "***communal discernment***", which assumes **a holistic understanding of the truth**.

> "*The truth which will emerge if all stakeholders generate a dialogue.*"
>
> (Cox 1986).

Dialogue spaces are especially important, but dialogue spaces where all the actors from the different social complex systems have a voice where we can establish a safe space to dialogue, where different paradigms, different points of view and perceptions can coexist.

Dialogue recognizes mutual humanity.

Mutual humanity refers to the recognition and understanding of our shared humanity, emphasizing the inherent dignity, worth, and interconnectedness of all people. It is a concept that promotes empathy, compassion, and cooperation among individuals and communities, regardless of their differences. Mutual humanity recognizes that despite our diverse backgrounds, cultures, beliefs, and experiences, we all share fundamental aspects of being human. It acknowledges that we have common needs, desires, and emotions, and that we all deserve respect, fairness, and equal opportunities. The concept of mutual humanity is often invoked in discussions about human rights, social justice, and global cooperation. It encourages us to transcend divisions and biases, promoting inclusivity, understanding, and solidarity among individuals and across societies.By embracing the idea of mutual humanity, we can foster a more harmonious and equitable world, where people work together to address shared challenges, promote human well-being, and create a better future for all.

Dialogue recognizes our differences, values our diversity, participants must be willing to show empathy towards others, recognize differences as well as areas of overlap, and demonstrate the capacity for change.

Dialogue is an Inclusive process.
Bring together a diverse set of voices to create a general pattern of understanding including discrepancies and conflicts.

Dialogue implies learning.
Modify the way people speak, think and how they communicate with each other. It generates a process of self-reflection that leads to personal change.

Dialogue addresses the root causes of a crisis and not just the symptoms.
It seeks to go beyond the negative peace defined by Galtung (1996) (Absence of Violence) and focuses on a positive peace that generates structural changes that lead to a genuine and sustainable peace.

Dialogue emphasizes a long-term perspective for change.
It seeks to find sustainable solutions that require time and patience, which can be slow and incremental.

1.4 Systems Thinking

Systems thinking is an approach and framework for understanding and analyzing complex phenomena as interconnected systems. It recognizes that individual components within a system are interdependent and that their interactions give rise to the behavior and properties of the system as a whole. Rather than focusing on isolated parts or events, systems thinking emphasizes the relationships and dynamics among these parts. (Navarro 2022)

Holistic Understanding of a problem as a system with interrelationships and changes.

"*Systems thinking is considered a much-needed competence to deal better with an increasingly interlinked and complex world. The many streams within systems science have diversified perspectives, theories, and methods, but have also complicated the field as a whole. This makes it difficult to understand and master the field.*" (Hieronymi 2013).

A comparative study (Wiek et al. 2011) came to the conclusion that one of the most important key competencies for a sustainable future is '*Systems thinking Competence'*.

Peter Senge, one of the key promoters of organizational learning and systems thinking in management (Senge 1990), argues that three core capabilities are necessary: We need to increase '*collaboration across boundaries',* 'see systems' as a part of larger systems and learn to '*create a desired future*' (Senge et al. 2010).

As Troncale states 'There is a need to make general systems theory more user friendly' (Troncale 1985).

1.5 Key Concepts of Systemic Thinking

Klir states that systems science, like any other science, needs to distinguish among three components: '*A domain of inquiry*', 'a body of knowledge regarding the domain' and '*a methodology*' (Klir 2001).

Key principles and concepts of systems thinking include:

Holistic Perspective: Systems thinking takes a holistic view of a system, considering the interactions, interconnections, and interdependencies among its components. It recognizes that changes or actions in one part of the system can have ripple effects on other parts.

Feedback Loops: Feedback loops are an essential aspect of systems thinking. They involve the flow of information or effects between different elements of a system. Feedback can be either reinforcing (positive feedback) or balancing (negative feedback), and it influences the behavior and stability of the system.

Emergence: Systems thinking acknowledges that systems can exhibit emergent properties or behaviors that arise from the interactions of their components. These emergent properties may not be directly attributable to individual parts of the system and can only be understood by studying the system as a whole.

Non-linearity: Systems thinking recognizes that system behavior is often nonlinear, meaning that small changes in one part of the system can lead to disproportionate effects or unexpected outcomes in other parts. Linear cause-and-effect relationships may not adequately explain the behavior of complex systems.

Boundaries and Perspectives: Systems thinking involves defining the boundaries of a system and understanding the perspectives and boundaries chosen for analysis. Recognizing the context and boundaries helps determine which components and relationships are included in the system and which are external to it.

1.6 Talking About Systems

By inducing systemic thinking in people, they begin to have a systemic perspective on the complex situations in which they are immersed. They begin to look at problems differently with a holistic vision, understanding that these problems are immersed in interconnected systems where various actors and variables permanently interact in a dynamic way.

The current state of the system variables leads the actors to make decisions, these decisions affect the variables, variables that are constantly changing. These dynamic interrelationships generate emergent properties of the system. Both properties and variables can be intervened if we understand inflection points.

When people understand that they are an important part of a larger system and that their actions have an impact on that system, they can better intervene in it. The best way to start these interventions is to understand the systems we are part of, their properties, and to be able to identify which variables we really have some control over, and to be able to focus our time and resources on those key variables. To do this we must talk about systems.

Having a common language and understanding that it is possible to act in complex situations makes people not feel powerless but rather empowered because they feel that their actions can contribute to change.

1.7 Systems in Practice

> *"The philosophers have only interpreted the world, in various ways; The point is to change it."*
>
> (Marx 1975).

Systems practitioners and academics have created a wide variety of responses in their attempts to tackle increasingly complex problem situations. Practitioners have increasibly bring together various systems ideas and techniques, methods, and methodologies in an organized way to employ them and improve particular problem situations. They are creating and using "*System Methodology Proposals*". The development of such methodologies started around the time of the second world war with the birth of Operational

Research, Systems Analysis, and systems Engineering. The combination of those methodologies were then labeled "*Hard Systems Thinking*" (Checkland 1981). Those practitioners assume that problem context is Simple and Unitary Systems. In the 60s and 70s these new proposals were taken to the universities and further refined by academics. Simple and Unitary assumptions assume that actors' stakeholders agree on objectives and that systems are enough to be modeled mathematically. Unfortunately, difficulties arise when those methodologies were used in more complicated and complex systems, systems that cannot be easily mathematically modeled or where traditional mathematical models will fail to capture the multiple interactions that exist between the parts both variables and actors, miss any emergent properties. Since 1960 serious attempts to extend the application of these methodologies have been made. This has allowed to extend the area of successful application of systems thinking methodologies by developing proposals that assumes problem contexts are more complex complicated pluralist and coercitives. Systems were as Kurtz & Snowden's describe complicated systems as systems were cause effect relationships are difficult to understand In Michael Jackson's Words;

"*This is the progress in applied systems thinking that we now seek to chart*" (Jackson 2019).

In the 70s and 80s Soft systems thinkers such as Churchman, Ackoff and Checkland developed their methodologies. They all focus on the complexity that arises from different stakeholders having a plurality of different beliefs and values acting according to different interpretations or perceptions of reality. Systems were there is not an agreement on goals or pretend to produce an objective. These new methodologies proposals try to generate between the actors systemic learning. A systemic learning process in which all the stakeholders come to appreciate alternative *Welstangshaung* or worldviews and the possibilities for change they offer, and reach a consensus or, at least, an accommodation about what needs to be done. Identifying what change is desirable and commit participants to agreed in course of action to be follow.

1.8 Cybernetics Organizational Systems

Cybernetics is a multidisciplinary field that explores the control and communication mechanisms in complex systems. It focuses on understanding how systems, including organizational systems, regulate and adapt themselves through feedback loops and information processing. In the context of organizational systems, cybernetics provides a framework for analyzing and improving the functioning of organizations by examining the flow of information, decision-making processes, and feedback loops within the system.

When evaluating the organizational structure of an organization understood as a complex social system (SSC), an important tool is Organizational Cybernetics. During the Second World War, interdisciplinary groups focused on creating solutions in complex social systems were created. These interdisciplinary working groups experienced unprecedented success and were known at the time as *"Operational Research Groups"*. Today more than ever, organizational problems transcend the scope of each discipline and cannot be addressed separately, but rather require a dialogue of knowledge and interdisciplinarity in order to offer real solutions to the needs and problems of an increasingly complex world. Since the end of the 1940s, the findings and developments of these groups extended to different fields other than the military, including engineering. One of these successful interdisciplinary groups was led by Professor "Norbert Wiener" and his experiences and findings gave rise to a new insight into the interactions of complex social systems and human responses to these interrelationships. Wiener was in charge of baptizing this new vision of complex systems as "Cybernetics" involving not only the work of his operational research group but also the work of other authors such as Warren McCulloch, Walter Pitts, Ross Ashby, Gray Walter among many others. Wiener defined this new science known as *"Cybernetics" as "the science of communication and control in the animal and the machine"* (Wiener 1948). Since its appearance at the end of the forties, multiple books and articles have been published on the subject, which has given rise to an international academic community interested in improving organizations understood as complex social systems. One of the most outstanding and recognized members of this community, Professor Stafford Beer, took a very important leap by proposing for the first time the concept of "*Organizational Cybernetics*" and the possibility of applying the learning and knowledge of operational research groups such as those of Norbert Wiener to Organizations (Beer 1959).

1.9 Viable Systems Methodology

The viable system model proposed by Stanford Beer sets out the organizational features that any system must possess if it is to be capable of responding to unforeseen environmental changes. The model is very general and easily applicable to organizations of all shapes. It seeks to explain how systems can survive and be effective over time by continuously self-organizing in the face of constant change.

Beer was the first academic to suggest applying the principles of the new science of cybernetics to the study of organizations understood as complex social systems. He strengthened the idea of the "*System*" and the organic structures opposed to the traditional mechanistic and reductionist approaches that were dominant in Western culture prior to the Second World War. He is also one of the first academics to defend the importance of self-regulating organic structures and holistic visions in the application of the scientific method and as Aristotle said in his time and more modern authors such as Aristotle (1045 BC) Capra (1998) or Navarro (2019) *"The whole is always more than the sum of the parts"* according to this idea the whole can only be fully understood if we can understand that its parts and the understanding of these in isolation differs and is inferior to nature of their sum. With this holistic, systemic and universal approach, complex social systems and organizations made up of people who are constantly making decisions that affect their environment can be analyzed.

Beer argues for the first time that by applying what he calls Organizational Cybernetics the innovative possibility of scientifically designing complex social systems and among these organizations such as the Salesian University Foundation. The way in which Beer proposes to achieve this is by providing complex social systems to organizations with "*basic capabilities of independent existence*" basic capabilities such as adaptation, flexibility, evolution, replication, reproduction, regulation, learning, among others. All these capacities are presented as necessary to be able to face the complexity of current needs and problems and the environments in which complex social systems operate. In his own words, Beer proposes that the systems be "*Viable systems*" (Pérez Ríos 2000–2008).

To ensure that a complex social system such as an organization meets the criteria established by Organizational Cybernetics in order to be considered not only an organic system but also a viable system, it must comply with a series of components. among which the 2 main ones are The Viable System Model and the Concept of Viability.

Let us start by defining the second: What does it mean for a complex social system, an organization to be a "*Viable System*".

As defined by Beer (1959, 1985) the Concept of Viability can be used to indicate the capacity of an organism understood as a system, an organization, a complex social system, to fulfill the "assigned purpose" to it, being able to ensure its survival. in time, survive in the face of changes that may occur in their environment and that may put their survival at risk. It is important to mention that many of these may not be foreseen or anticipated. It is in this sense that Beer suggests that Cybernetics and particularly Organizational Cybernetics can ensure the viability of complex social systems such as organizations (Beer 1985).

1.10 System Dynamics

System dynamics is an interdisciplinary approach and modeling methodology for understanding and analyzing the behavior of complex systems over time. It was developed by Jay Forrester in the 1950s and has been widely applied in fields such as engineering, economics, management, public policy, and environmental studies. System dynamics focuses on understanding the feedback processes, delays, and nonlinear relationships within a system. It aims to capture the dynamic behavior and interactions of variables over time to gain insights into system behavior, identify key leverage points, and inform decision-making. MIT Professor Jay Forrester proposed the systems dynamic approach taking advantage of the power of new computers studying levels and flows and providing the expertise necessary to gain insight into complicated and complex contexts. System Dynamics allow decision makers to uncover the structural relationships between the feedback loops that governs systems behavior. Once the relations are better understood, it becomes possible to decision makers to determine the leverage points that can be exploited to change the behavior of a particular variable by changing the behavior of positive and negative feedback loops to produce desirable outcomes.

1.11 Critical Systemic Thinking

Critical Systems Thinking (CST) is an approach that combines systems thinking with critical thinking to analyze and address complex problems. It aims to promote social transformation, justice, and sustainability by questioning and challenging the assumptions, values, and power dynamics embedded within systems.

As Luhman states (2013) "If one introduces an observer, a speaker, or simply one to whom one attributes a statement, one relativizes ontology… One always faces the questions of who says a particular thing and who does something and from which system perspective the world is seen in a particular way (and no other)".

Many systems thinkers guide their job based on the assumption that "The hole is more than the sum of its parts and that everything is connected" But this approach is criticized by others like Bryer (1980) and Phillips (1976) who say that systems thinking is hopelessly idealistic and unworkable, because you are supposed to stary with an understanding of the whole system because in a way "you need to know everything to know anything". And unless you really know the whole system you cannot justify taking any action to intervene the system because you are never going to be able to really anticipate the results. But as Ulrich put this "We cannot blame the messenger for the bad news" (Ulrich 1981) and it make us understand that even if the understanding of the whole is a theoretical ideal that makes us reflect on the nature and accuracy of our knowledge and the fact that we don t comprehend a 100% of the systems we work on, it is actually possible to work with limited conceptions of the whole as long as we are transparent about what we are doing and sharing the limitations of our knowledge about the system and the interventions.

We need to find a process so that we can proceed knowing that it is impossible to achieve a full comprehensive knowledge of the system we are working with and for this we need to implement the "Critical Systems Thinking".

> A systems of systems methodology is developed as the interrelationship between different methodologies is examined along with their relative efficacy in solving problems in various real world problem contexts…The analysis points to need for a coordinated research programmed designed to deepen understanding of different problem contexts and the type of problem-solving methodology appropriate to each
>
> (Jackson and Keys 1984, p. 473)

> Critical systems thinking offers an alternative approach to validating work in social systems science. This is a critical approach that derives from the work of an author called Jurgen Habermas wich as its claimed by several other authors builds on the strengths and overcomes all the weaknesses of both the hard and soft versions
>
> (Jackson 1982)

"The Starting point for a project should be what the clients is actually interested in" (Jackson 2019).

The ideas that inspired the development of this new Critical Systems Theory, they derive from two main sources.

1. The Social Theory
2. The Systems Approach.

From the social science side work as for example the work of Burrell and Morgan's (1979) on social Paradigms and organizational analysis shows the particular importance on having different ways of analyzing and intervening in organizations. This work support the critique of assumptions that different systems methodologies and approaches make about the particular social realities they are working with or even the social science itself.

Critical social theory from social science referents as Foucault, Marx or Habermas, had a significant influence in the development of CST.

From **Systems Theory** and the systems approach, CST has inherited important key characteristics as the holism and a large and very powerful set of concepts, methodologies & methods based upon those key concepts for intervening systems and redesign them looking for improvement of problem situations. CST is developed by mixing both of the strengths of social theory and systems thinking. On the one hand the social sciences are strong on theory, on thinking about the ontological and epistemological assumptions that are used to the intention of creating or gaining new knowledge and to use it in practice. Social Sciences combined with systems thinking can provide important means for reflecting on possible community interventions and strongly improving the methodologies or methods that are used. On the other hand, systems thinking is able to translate all the findings from social theory into a practical form to include them in well-developed methodologies to use them to redesign and to create the desired change in the intervened systems. The problems that we face when we intervene complex social systems, the problems of the real world rarely correspond to a traditional disciplinary boundary and the systems thinking and the systemic view and ideas always encourage transdisciplinary and multidisciplinary approaches.

1.12 Complexity Theory

Complexity theory, is a multidisciplinary framework that seeks to understand and explain the behavior of complex systems. It explores how diverse elements within a system interact and give rise to emergent properties, patterns, and behaviors that cannot be easily predicted or understood by examining the individual components alone.

For Gleick and Berry (1987) "Chaos theory" is the forerunner of "complexity theory" and complexity theory is one of the three landmark achievements of the twentieth century science (). Mathematician Poincaré (1854–1912) is consider the founding father of chaos and complexity theory. He tried to solve the so-called "Many-body problem" in celestial mechanism according which Newton's Laws were fine when the behavior of two bodies exerting gravitational forces on one another wanted to be predicted. But it was almost impossible to do the same with three bodies subject to mutual attraction. Henry Poincaré didn't manage to find a solution but he manages to describe the problem he was struggling with.

> ...but even if it were the case that the natural laws had no longer any secret to us, we could still only know the initial situation approximately. If that enabled us to predict the succeeding situation with the same approximation, that is all we require, and we should say that the phenomenon has been predicted, that it is goberned by laws. But it is not always so; it may happen that small differences in the initial conditions produce very great ones in the final phenomenon. A small error in the former will produce an enormous error in the latter. Prediction becomes impossible...
>
> (Poincaré–Mitchell 2009).

Sensitive dependence on initial conditions is one of the signatures of chaos theory.

Another important contribution was made by Darwin back in 1859 with his book the origin of species. His theory stated that random variation in the traits possessed by individual organisms, passed on by inheritance, can lead to a particular species flourishing because the new traits are well adapted to the environment. In this case evolution proceeds not on the basis of determinism but as a result of natural selection of the outcomes of chance events. Defying Entropy the second law of thermodynamics living systems are able to become more complex and intricate. In words of Charles Pierce (Boulton et al. 2015).

> Now the only possible way of accounting for the laws of nature and the uniformity in general is to suppose them results of evolution. This supposes them not to be absolute, not to be obeyed precisely. It makes an element of indeterminacy, spontaneity, or absolute chance in nature
>
> (Boulton et al. 2015).

1.13 Complexity Theory and Systems Thinking

Complexity theory and systems thinking are closely related and interconnected. They both offer frameworks and perspectives for understanding and analyzing complex phenomena and systems. While they have some overlapping principles and concepts, they approach complexity from slightly different angles.

In essence, complexity theory can be seen as a subset or application of systems thinking, focusing specifically on the behavior and properties of complex systems. Complexity theory builds upon the foundation of systems thinking but provides a more specialized lens for understanding the unique characteristics of complex systems. By combining systems thinking and complexity theory, practitioners can gain a more comprehensive understanding of complex phenomena and systems. They can explore the interconnectedness, feedback dynamics, emergent properties, and adaptive behaviors of systems, which can inform decision-making, problem-solving, and intervention strategies in a wide range of domains.

Stacey et al. (2000) states that Systems thinking has its limitations because of the limited ways of understanding organizational life. But they are new methodology proposal that offer some alternatives enabling Complexity Theory to find a voice and to coexist. Stacey's regards system's thinking as being the main element in the dominant discourse on management. It required a drawing of boundaries leaving elements outside the system. This encourages an approach to management in organizations based on two different ideas.

The first one is that managers are seeing in rationalist way as if they could manage the organizations they are responsible for from the outside using a sort of rational means to control and oversee the interaction of all the different elements looking to improve their behavior.

The second one is that organizations are presented as systems governed by a formative approach which sees them as unfolding their potential in a predictable way. But there for human purpose is excluded from the organization itself which bring some problems. This idea will lead to create in managers an illusion that they are in full control of the organization but things will not happened as planed and they will be frustrated and constantly disappointed when thing go in different directions than expected. Freedom and agency of people in the organization its important it brings creativity innovation and novelty and the self-organization that comes from the creation of relations and their interaction. We all know that inside an organization diversity of perspectives, conflicts and change occur.

> "*The dominant discourse takes its own ideology for granted and through its mainstream theories largely sidelines matters to do with interdependence, unpredictability, power, ordinary day-to-day activities such as conversation and politics, and the richer fabric of social life.*"
>
> (Stacey and Mowles 2016).

1.14 The Value of Systemic Thinking and Complexity Management to Decision Makers

Complexity Management: Complexity management refers to the strategies, approaches, and techniques used to understand, navigate, and effectively deal with complex systems or situations. It involves recognizing and addressing the challenges that arise from the interdependencies, uncertainties, and non-linear dynamics inherent in complex systems. Complexity management acknowledges that many real-world problems and systems are characterized by a high degree of complexity, which can make them difficult to understand, predict, and control using traditional linear approaches. Complexity can emerge from factors such as multiple interacting components, feedback loops, non-linear relationships, and dynamic behaviors.

Systemic thinking and complexity management offer significant value to decision-makers in various ways:

Understanding Interconnections: Systemic thinking helps decision-makers recognize and understand the interconnections and interdependencies within complex systems. It enables them to see how different elements within a system interact and how changes in one part can impact other parts. This understanding helps in anticipating and managing potential ripple effects and unintended consequences of decisions.

Holistic Perspective: Systemic thinking encourages decision-makers to take a holistic view of problems and situations. Instead of focusing on isolated components or symptoms, they consider the entire system and its dynamics. This broader perspective allows decision-makers to identify systemic root causes, develop comprehensive solutions, and avoid narrow, short-term thinking.

Identifying Leverage Points: Systemic thinking helps decision-makers identify key leverage points within a system, which are areas where small changes can have significant and lasting impacts. By focusing on these leverage points, decision-makers can prioritize their efforts and resources for maximum effectiveness.

Anticipating and Managing Complexity: Complexity management equips decision-makers with tools and approaches to understand and navigate complexity. It helps them make sense of complex situations, analyze nonlinear relationships, and account for uncertainties and dynamic behaviors. This enables decision-makers to make more informed decisions and develop strategies that consider the intricacies of complex systems.

Embracing Adaptability and Resilience: Systemic thinking and complexity management emphasize the importance of adaptability and resilience in decision-making. Decision-makers understand that systems are dynamic and subject to change, and they need to develop strategies that can adapt to evolving circumstances. They consider multiple scenarios, explore alternative futures, and build resilience to cope with unexpected events.

Considering Multiple Perspectives: Systemic thinking encourages decision-makers to engage with diverse stakeholders and incorporate multiple perspectives. By actively seeking different viewpoints, decision-makers can gain a more comprehensive understanding of the system and its complexities. This inclusive approach can lead to more robust and inclusive decision-making processes.

Addressing Systemic Challenges: Many of today's challenges, such as climate change, social inequalities, or global pandemics, are complex and systemic in nature. Systemic thinking and complexity management offer decision-makers the tools to tackle these challenges effectively. They help identify interconnected causes, develop holistic solutions, and promote long-term sustainable outcomes. By applying systemic thinking and complexity management, decision-makers can make more informed and effective decisions. They can navigate complex systems, anticipate unintended consequences, and develop strategies that address the root causes of problems. This ultimately leads to more sustainable, resilient, and impactful outcomes in a rapidly changing world.

The Managerial world is random, unpredictable, messy. Complexity Theory offers a better description of managers world than other management theories.

It also can help to find order among Chaos. In a way the insights of complexity theory can been of great importance for improving management theory and practice.

They are more and more successful application of Complexity Theory and Systems Thinking that demonstrate that both Systems Thinking and complexity theory affects positively management style, structure and process.

> "*The natural scientific domain of complexity theory is better understood than that of management. There are concepts of the natural domain of complexity that can clearly put in one-on-one correspondence with similarly precise equivalents in the domain of management. Connections between groups of concepts in the first domain are implicitly preserved between their equivalents in the second*"
>
> (Rosenhead 1998).

It is at this meeting point between Complexity Theory and Systems Thinking that we can find relevant tools that can help decision makers in organizations to better navigate complex social systems (CSS) improving their understanding of the systems of which they are a part helping them to optimize limited resources and accelerating the implementation and materialization of projects.

1.15 Community Development

Community development refers to a process aimed at enhancing the social, economic, and cultural well-being of a specific community or group of individuals. It involves empowering community members to participate in identifying and addressing their needs, aspirations, and challenges, ultimately leading to positive change and sustainable development.

References

Ackoff RL (1999a) Ackoff's Best : His Classic Writtings on Management New York :Wiley.
Ackoff, Russell L. Re-creating the corporation: A design of organizations for the 21st century. New York: Oxford University Press, (1999b).
Ackoff, RL, Gharajedaghi (1996) Reflections on systems and their models
Aristotle (1045 BC) Metaphysics 8–10
Beer, S. (1959). Cybernetics and Management. English Universities Press, Oxford.
Beer, S. Diagnosing the System for Organizations.John Wiley & Sons. (1985).
Boulton, J, Allen, P, Bowman, C (2015) Embracing Complexity :Strategic Perspectives for an Age of Turbulence Oxford :Oxford University.
Bryer, RA (1980) some comments on Churchman and Ulrich Reply to the status of the systems approach.
Burrell, G and Morgan, G (1979) Sociological Paradigms and Organizational Analysis. London: Heinemann.
Capra, F., Sempau, D. (1998). The Web of life
Cox, G. (1986). The Ways of Peace: A Philosophy of Peace as Action. Mahwah: Paulist Press.
Galtung, J. (1996). Peace by peaceful means: peace and conflict, development and civilization. Thousand Oaks: Sage.
Goldstein, J. S., & Freeman, J. R. (1990). Three-way Street: Strategic reciprocity in world politics.
Gibbons, M., Limoges, C., Nowotny, H. et al. (1994). The New Production of Knowledge:The Dynamics of Science and Research in Contemporary Societies. London :Sage.
Gleick J, Berry M (1987) Chaos Making a New Science
Hall & Fallen (1956) General Systems Theory
Hieronymi. A, (2013) Understanding Systems Science: AVisual and Integrative Approach University of ST Gallen Switzerland 2013 The Author. Systems Research and Behavioral.
Heylighen F. (1990). Classical and non-classical representations in physics I. Cybernetics and Systems 21: 423–444.
Holland JH. (2006). Studying complex adaptive systems. Journal of Systems Science and Complexity 191.
Jackson, MC (1982) The nature of Soft Systems thinking: The work of Churchman, Ackoff and Checkland. Journal of Applied Systems Analysis 9:17–28
Jackson, M.C and Keys, P (1984) towards a system of systems methodologies. Journal of the operational research society 35:473-486.
Jackson M (2019) Critical Systems Thinking and the Management of Complexity. Responsible leadership for a complex world.
Jervis, R (1997) Complexity and the analysis of political and social life. Political science quarterly, 1997—JSTOR
Klir, GJ. (2001). Facets of Systems Science. Kluwer: New York.
Kelso, J. A. S. (1995). Dynamic patterns: The self-organization of brain and behavior.
Luhman N. (2013) Introduction to Systems Theory Cambridge Polity Press.
Marx K (1975) from 1844 Manuscript Economic and philosophical manuscrpts. In Marx: Early Writings, 279–400. Harmondsworth:Penguin EDT.

Meadows DH (2004) Limits of Growth : The 30-year update.
Mitchell,M (2009) Complexity :A guided Tour .Oxford Oxford University.
M'Pherson PK. (1974). A perspective on systems science and systems philosophy. Futures 6: 219–239.
Navarro, J. (2001). "Las organizaciones como sistemas abiertos alejados del equilibrio". Tesis de doctorado no publicada. Universidad de Barcelona: España.
Navarro, CA (2019) The Whole is always more than the sum of the parts—The Complexity Funnel & The Social Transformation Workshops: A Model for Sustainable, Holistic Change-Global Education Leadership Organization-Activating leadership to help eradicate poverty
Navarro C, (2020) Social Complex Systems (SCS) Co-creation of a systemic model to support community engagement projects.
Navarro, CA (2022) Untangling Complexity Complexity Funnel Methodology & Social Transformation Workshops Universidad de los Andes Bogotá DC Colombia
Pérez Ríos, J. Diseño y diagnóstico de organizaciones viables. Un enfoque sistémico. Iberfora (2000–2008).
Phillips, D.C. (1976) Holistic Thought in Social Sciences . Stanford, CA: Stanford.
Rosenhead, J (1998) Complexity theory and management practice. London. London School of Economics LSE Working Paper, 98.25.
Senge P. (1990). The Fifth Discipline. The Art and Practice of the Learning Organization. Doubleday/Currency: New York.
Senge P, Smith B, Kruschwitz N, Laur J, Schley S. (2010). The Necessary Revolution: Working Together to Create a Sustainable World. Broadway Books: New York.
Stacey, RD, Griffin D, Shaw, P (2000) Complexity and management: Fad or radical challenge to systems thinking?
Stacey RD & Mowles, C (2016) Strategic Management and Organizational Dynamics the Challenge of Complexity Two Ways of Thinking about Organizations. 7e Harlow Pearson.
Strijbos S. (2010). Systems thinking. In The Oxford Handbook of Interdisciplinarity, Frodeman R, Thompson KJ, Mitcham C. (Eds.). Oxford University: Oxford; 453–470.
Troncale L. (1985). The future of general systems research: obstacles, potentials, case studies. Systems Research 2: 43–84.
Tranfield, D. and Starkey K (1998).The nature, social organisation and promotion of management research:towards policy. Brtish Journal of Management 9:341–355.
Ulrich, W (1981) On blaming the messenger for the bad news : Reply to Byers Comments Omega 9 200–202.
Vesilind, A. (2010). Engineering Peace and Justice. London: Springer books.
Webel, C. y Galtung, J. (eds.). (2007a). Handbook of Peace and Conflict Studies. London: Routledge.
Webel, C. y Galtung, J.(2007b) Manual de Estudios de Paz y Conflicto. Londres: Rortledge.
Wiek A, Withy combe L, Redman CL. (2011). Key competencies in sustainability–a reference framework for academic program development. Sustainability Science 62: 203–218.

Complexity Funnel Methodology 2

2.1 Development

Complexity Funnel Methodology Proposal:

For the past six years work has been done on the development of a pluralistic multi-methodology proposal to help decision makers in social complex systems (SCS) make better decisions.

Better decisions understood as decisions that help to optimize the use of limited resources and accelerate the implementation of projects but at the same time being legitimized by the actors that are part of the systems, thus serving as a catalyst on community projects implementation.

This multi-methodology, titled the "*Complexity Funnel Methodology*" (CFM), takes the form of a series of "*Social Transformation Workshops*" (STWs) that seek to map the current state of the SCS, with emphasis placed on the identification of main elements, actors and variables, the interrelations between them, and the resulting impact of the relationships on various variables, possible obstructions both discrepancies and conflicts that may exist within those systems and create social complexity and the emergent properties of those systems. The goal is then to reduce the social complexity within the system systematically applying the STWs.

Justification:
Minger's vision of Critical Systems Thinking and Multi-methodologies:

> "*The fundamental tenet of multi-methodology is the importance of combining together methods or sometimes methodologies in dealing with real world situations, whether the purpose is pure research or a practical intervention or both...*

C. A. Forero Navarro, *Untangling Complexity—Peace Building Engineering*,
Synthesis Lectures on Engineers, Technology, & Society 29,
https://doi.org/10.1007/978-3-031-37038-0_2

> *The first argument in favor of multi-methodology is that the real world is complex and multi-dimensional, while particular research or intervention methodologies focus only on specific aspects.*
>
> *The second is that research and intervention is not a discrete event but a process that has phases, different types of activities that will predominate at different times.*
>
> *A third argument is that it encourages triangulation of research and generates more interesting and stimulating results.*"
>
> (Mingers 2014).

The idea of using multi-methodologies (MM) means:

> "*To use a combination of methodologies from different paradigms and methods together in a single intervention*"
>
> (Jackson 2019).

The growing interest in MM and pluralism started in the early 80s the first authors talking about MM in critical systems thinking were Jackson and Keys (1984); Jackson (1987).

This was the starting point of theoretical work in critical systems thinking (CST) and multi-method applications by consultants in different types of complex social systems (CSS). This is also the starting point of a growing academic community interested in MM and pluralism.

Pluralism started to become necessary judging for the number of reported interventions using MM (Munro and Mingers 2002) it seemed like combining methods brought success in every intervention.

More recently in 2010 a review on the influence of Systems Thinking in OR (Mingers and White 2010) such the diversity and success on the application of MM CST in management, quality management, designing and support areas.

They are many other more recent Applications, like Henao and Franco (2016; Acero (2021) community engagement in Colombia or Zlatanovic (2017) Strategy formulation and Planning in Serbia, Chowhury and Jangle (2018) community engagement in India, Navarro and Dalrymple (2019) community engagement Us-Mexico border.

Jackson collects 5 main reasons why multi-methodology thinking and pluralism are desirable and more and more popular.

When he talks about popularity, Jackson refers to the number of academics implementing mixed or multi-methodological approaches in their work around the world.

Below we explain in detail the 5 reasons why Jackson affirms that pluralism and multimethodologies are desirable!

1. The world we live in, and we confront and seek to transform by our interventions is "Multidimensional" (Mingers and Brocklesby 1997). Argues like Habermas that any real-world situation is a complex interaction of substantively different elements.

   "There will be aspects that are relatively hard and observer independent, particularly material, and physical processes, that we can observe and model. There will be aspects that are socially constituted, dependent on particular cultures, social practices, languages, and power structures, that we must come to share and participate in. Finally, there will be aspects that are individual such as beliefs, values, fears, and emotions that we must try to express and understand."

   (Mingers 1997a).

   Meaning that different paradigms will focus on different aspects of a particular problem, highlighting some features and hiding others. It makes sense working with multiple paradigms looking through a variety of theoretical lenses employing different generic CST methodologies trying to capture as many different facets of a problem as possible and plan an intervention in a more informed and comprehensive manner.
2. Research and Intervention are processes that are likely to be organized and proceed in a number of phases. Each phase will then involve different tasks and pose different problems for researchers and practitioners. With this in mind it is a good thing to have a wide spectrum of methodologies and methods available to face this problems situations.
3. Multi-methodology Thinking and pluralism sounded like a post-modernist argument against gran narratives.

   "*.... Simplifying to the extreme, I define postmodern as incredulity towards metanarratives... Postmodern knowledge is not simply a tool of the authorities, it refines our sensitivity to differences and reinforces our ability to tolerate the incommensurable. Its principle is not the expert's homology, but the inventor's paralogy.*"

   (Lyotard 1984).

4. Lack of success of traditional methodologies and methods to deliver promised results.

   Is not a secret that hard systems approach was not successful in practice when confronted with the complexity of social systems and strategic problems, it just couldnt deliver the promised results, or bring any competitive advantage to the organizations were they were implemented. This is the reason why academics start questioning the old ways and made important advances in new approaches giving birth to the pluralistic and multi-methodology approach.

5. Mingers states that MM allows a form of triangulation, were a decision maker can use different lines of sight to accurately fix a position. If a similar result is obtain by using different research approaches then researchers will have more confidence on the validity of their findings.

> Looking a phenomenon from various perspectives provides a greater appreciation of its complexities, especially if what we see differs.
>
> (Mingers 1997b).

The Pluralist or MM approach opposes the "isolationist" traditional approach. Accepting that different theoretical and methodological approaches are legitimate and are able to address problems situations differently is a first step towards pluralism.

> "*Recognizing the need to use different approaches in combination to tackle one problem situation because of its multidimensional characteristics that's the real breakthrough to fully fledge pluralism.*"
>
> (Jackson 2019).

Off course depending on the specifics of each intervention it might be important to alter the methods that were chosen during each phase of the intervention. And here there is very valid concern raised by Eden who states that.

> "*This will impose to decision makers to know or to be advice about how to combine approaches that may rest on different social theories.*"
>
> (Eden et al. 2009).

In Michael Jackson´s words.

> The complexity, heterogeneity and turbulence of many problem situations demand that pluralism must be exercised at three levels
>
> 1. Methods
> 2. Methodologies
> 3. Paradigms.
>
> (Jackson 2001, 2003, 2019).

A Pluralistic Multi-methodology

A pluralistic multi-methodology approach refers to an inclusive and flexible approach that integrates multiple research methods and approaches to gain a comprehensive understanding of a particular phenomenon or research question. It recognizes that different methods have their strengths, limitations, and areas of applicability, and seeks to combine them to leverage their respective advantages.

Pluralistic Multi-methodologies Approach look at problems situations as messes that cannot be understood from only one point of view but from a variety of perspectives. Obstructions will appear both discrepancies and conflicts but once an agreement is reached between the actors about the existence of these obstructions they can start working together to solve it. There will be a common model shared by the actors where the obstructions will be highlighted, there will be agreements about the mayor issues and problems they are facing as part of the same system. The choice of methods and methodologies to address and manage this mess and tackling the main issues must be made in full knowledge of the strengths and weaknesses of the different systems approach.

Michael Jackson thoughts about this combination of methods and methodologies are as follow:

> "*The idea of naming* **dominant and dependent** *methodologies and switching between them helps with the practical difficulties associated with multi-paradigm working, problems that arise can be manage if an initial choice of dominant methodology is made, to run the intervention, with a* **dependent methodology** *or methodologies reflecting alternative paradigms in the background. The relation between dominant and dependent methodologies can then change as the intervention proceeds in order to maintain flexibility at the methodology level...*
>
> *It allows the intervention to proceed in a theoretically informed way, making research possible, and protects paradigm diversity while being less confusing to participants. As long as we are explicit about our initially* **dominant methodology** , *and are ready to switch as necessary, proceeding initially with one methodology is not damaging. That initial choice does not exclude us from introducing* **alternative methodologies**, *based on different paradigms, as the situation demands*"
>
> (Jackson 2019).

Jackson used the philosopher Althusser to justify his strategy of alternating dominant and dependent methodologies.

Althusser conceived the social totality.

> "*... As a system that at various times is dominated by one of its main instances like economic or political or theoretical or ideological... among many others. In order to be able to change social totality decision makers should be able to understand the interrelations that exist between these main instances how they develop individually and how the dominance of one or another changes over time.*"
>
> (Althusser et al. 1970).

Jackson replaces the notion of dominant instances presented by Althusser and replacing it with the idea of improving efficiency viability mutual understanding fairness among others.. Which becomes the dominant concern in any problem situation for some ´period of time (Jackson 2000).

With the right amount of time, they will eventually all be addressed but is just impossible to do it all at once.

Pluralistic multi-methodologies is an iterative and systemic process it privilege those aspects that need to be addressed more urgently but it never stops is a permanent continuous improvement cycle in a complex and turbulent environment.

The problem is not static the problem is dynamic and how the problem is perceived by those who are part of the social complex system changes too, new intervention strategies need to be planned the only way to successfully do it is;

> "… *To continually cycle around Creativity, Choice and Implementation, changing methodologies that are dominant and dependent.*"
>
> (Jackson 2019).

This approach suggested by Jackson encourage highly creative thinking to really understand the unique characteristics of each social complex systems and the problem situations that exist within. Obstructions, discrepancies, and conflicts.

> "*In highly complex problem situations it's advisable to address at the same time different aspects by taking into account different perspectives. To do so decision makers can use different systems methodologies in combination but having one as dominant and the others as supportive.*"
>
> (Flood and Jackson 1991).

There are those that doesn't agree with Flood and Jackson, Gerald Midgley in his book "*What is This Thing called Critical Systems Thinking*" (Midgley 1996a, b) says that in his view Jackson Total Systems Intervention is not meta-paradigmatic. Because in Midgley words

> "…. *TSI embodies its own unique assumptions with which other theorists as him do not agree…related with Habermas theory of knowledge constitutive interest this theory rest on an untenable belief in the march of progress towards universal emancipation and risk ignoring the need for sustainable improvement that takes the natural environment into account. He also agreed with Ulrich that TSI has a restricted conception of the role of critique and only seemed to bring it into play when coercive contexts are identified…*
>
> *Critique is needed all time not least to identify cases of coercion… This is specially the case when working for a specific client when they are pressures to take organizational boundaries for granted and not to consider the impact of interventions on other stakeholders.*"
>
> (Midgley 1996a, b).

But Midgley made a constructive criticism of TSI he suggested a way to solve those difficulties using the work of Churchman and Ulrich.

Churchman States that

> "...*boundaries drawn around a system determine what is taken into account, which stakeholders are considered, how improvement is defined and what actions are contemplated.*"
>
> (Churchman 1979a, b, c).

He also argues that the boundaries of systems are not fixed by the structure of reality but by the knowledge and values of observers and so that in order to justify complex systems interventions we need to take in to account several different perspectives of a wide variety of stakeholders giving special attention to those that tend to be excluded or that that doesn't t have a voice.

Ulrich took this idea from Churchman and the knowledge derived from Habermas work on discursive rationality and created Critical Systems Heuristic.

Ulrich methodology offers a means of supporting rational professional practice that depends neither on the impossible sweeping in of all stakeholders nor the utopian vision of an ideal speech situation. Critical Systems Heuristic insist that appropriate boundary judgments can only be established through dialogue between those involved in designing the systems and those who are affected by the design. It also provides means to be able to detail the nature of different boundary judgments and allowing them to be rigorously debated.

Midgley join Churchman and Ulrich in their proposal without losing those aspects of methodological pluralism which he values and feels are legitimate.

This leads a Midgley to propose a reconstruction of Critical Systems thinking called SI Systems Intervention.

Midgley defines Systems Interventions SI as.

Systems Interventions

> "*A purposeful action by an agent to create change in relation to reflection upon boundaries.*"
>
> (Midgley 2014).

His methodology proposal starts with **a reflection upon boundary judgments.**

Then second involved "**Creative Design Methods**" it makes use of the variety of available methods found in the systems literature and even beyond including methods from other areas to be able to create a flexible and responsible approach to intervention.

About the first step.

A reflection upon boundary judgments

> "*...boundary critique must not be subordinated to methodology choice but needs to preceed it. The separation of the two methodological concerns of critique and emancipation is not tenable on critical grounds.*"
>
> (Ulrich 2003).

Following Ulrich thought Midgley suggest starting by asking questions about boundary critique in any intervention overcoming the problems presented in TSI, He argues that a new paradigm can be built around the ideas that appears with the boundary exercise. Wich is very suited to CST interventions. Boundary critique enables decision makers to overcome subject-object dualism one of the most important enemy of systems thinking and one that has so far not been dealt with just yet!

In Midgles words

> The problem of subject-object dualism can be overcome if we make boundary judgments the central piece, the corner stone of the intervention process.
>
> "*...if we regard the process of making boundary judgments as analytically prime, rather than a particular kind of knowledge generating system, then subjects come to be defined in exactly the same way as objects by a boundary judgment.*"
>
> (Midgley 2000).

Following Midgley's thought critical systems thinking shouldn´t continue making the unsustainable claim that it is multi-paradigmatic. Because he can proudly say:

> "*... That it has its own process paradigm based on boundary judgments. This paradigm will be subject to attack from non-believers especially those who continue to adhere to various forms of dualism and so CST is again subject to paradigm disputes.*"
>
> (Midgley 2000).
>
> "*the exercise of process philosophy... allow us to talk about the real world social construction and subjective understanding without contradicting ourselves... it is preferable to realism, idealism and social constructions because it can offer an alternative grounding for most of the work from these other paradigms without the need to alter their most important contributions.*"
>
> (Midgley 2000).

This approach has been used in projects for example (Cohen and Midgley 1994) and also Boyd et al. (1999); Navarro and Dalrymple (2019)

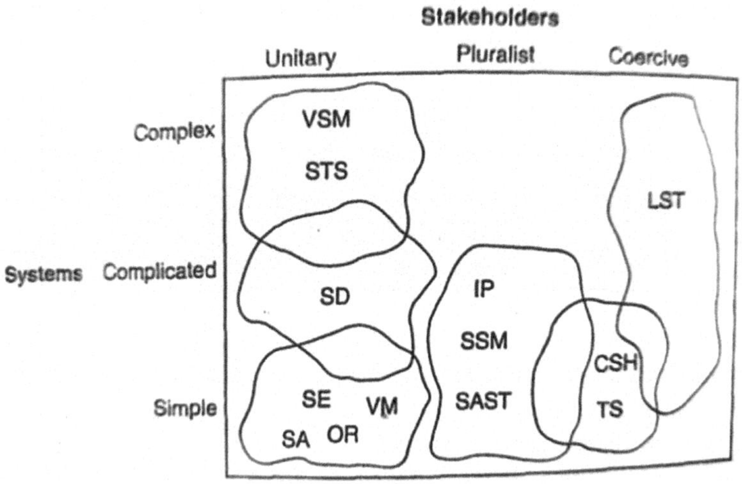

Fig. 2.1 Position of all critical systems thinking methodologies on the systems of systems methodology SOSM

Complexity Funnel Methodology Proposal:

Following Habermas, Churchman, Michael Jackson, and Gerald Midgley´s more than 50 years reflections and discussions upon pluralistic multi-methodologies we develop our own Complexity Funnel Methodology Proposal. The Complexity Funnel Methodology Proposal (Fig. 2.1).

Inspired by Jacksons Position of all critical Systems thinking methodologies we create a Spectrum of Possible methodologies that could be used by the decision makers of the social complex systems we were working with.

For 6 years we tested the impact of different methods and methodologies to find out which were the most effective to support the implementation and materialization of projects in both private mixed and public companies, communities at a national and international level and peacebuilding communities (Table 2.1).

Decision makers in complex social systems were presented with different critical systems thinking tools that could be used to generate a better understanding of the system and to help them make decisions. At the end of the interventions, they were asked to rate the relevance and impact of the different methodologies used to know which were the ones that facilitated their understanding of the system of existing obstructions, both discrepancies and conflicts, and that helped them make better decisions or more informed decisions.

Results of interventions in private, public companies, national and international communities' shows that some classic Systems thinking methodologies and methods are perceived by decision makers as responsible for acceleration materialization of their

Table 2.1 Perception of relevance of the methodologies as a catalyst of processes in companies according to decision makers between 2016–2021

Metodologies/SCS	Private companies		Public Companies		National Comunities	International Comunities		
Metodologies/Interventions	Henkel	Merqueo	Colombian Republic Bank	Cartagena Maire	Venezulean Migrants	Mexican Students	Total	Priorization
IP	2	2	2	2	2	2	12	1
SSM	2	2	2	2	2	2	12	1
CSH	2	2	2	2	2	2	12	1
VSM	2	2	2	1	2	2	11	2
OR	2	1	2	1	2	2	10	3
TS	2	1	2	1	2	2	10	3
SD	1	2	2	2	1	1	9	4
STS	1	1	2	1	1	1	7	5
SE	1	2	1	1	2	1	8	5
SA	1	2	1	1	2	1	8	5
SAST	1	1	2	1	1	1	7	5
LST	1	1	2	1	1	1	7	5
VM	1	1	1	1	1	1	6	6
Total	19	20	23	17	21	19		

projects, we decided then to privilege those methods and methodologies in Peace Building communities' interventions.

Among those methodologies we can find

- Systems of Systems Methodology Peter Checkland
- VSM Viable Systems Methodology Stanford Beer
- Organizational Pathologies
- Interactive Planning and Idealized Design
- Critical Systems Heuristic Wegner Ulrich
- Systems Dynamics Sterman.

An explanation of the methodologies its pros and cons and its strengths and weaknesses were shared with the decision makers of the intervened systems.

Knowing and understanding the methodologies, their methods, and the expected results of their applications in the complex social systems in question, an action plan and a clear timetable for the application of some workshops were established with the decision makers.

Each intervention was unique and flexible it was adaptable and could easily change depending on the monitoring and control process because social complex systems themselves change over time they are dynamic not static. They require processes of automatic reflection, monitoring and control that allow continuity of interventions but adapting to changes (Table 2.2).

These workshops have previously been tested and refined based on implementations in private, public, mixed companies, communities at a national and international level and more recently in Peace Building communities in Colombia (Figs. 2.2, 2.3, 2.4 and 2.5).

This new methodology and its application are grounded in works of other systemic thinkers and scholars such as Midgley (2000) and Rosenhead (1986).

When pluralistic multi-method approaches are utilized in a community setting, vs. industrial setting, (Howick and Ackerman 2011) mention the need for further investigation and development. Howick and Ackerman studied more than thirty multi-methods

Table 2.2 Types of systems Michael Jackson (2019) critical systemic thinking and the management of complexity

Systems	Unitarian	Pluralistic	Coercive
Simple	Simple unitarian	Simple pluralistic	Simple coercitive
Complicated	Comlicated unitarian	Complicated pluralistic	Complicated coercive
Complex	Complex unitarian	Complex pluralistic	Complex coercitive

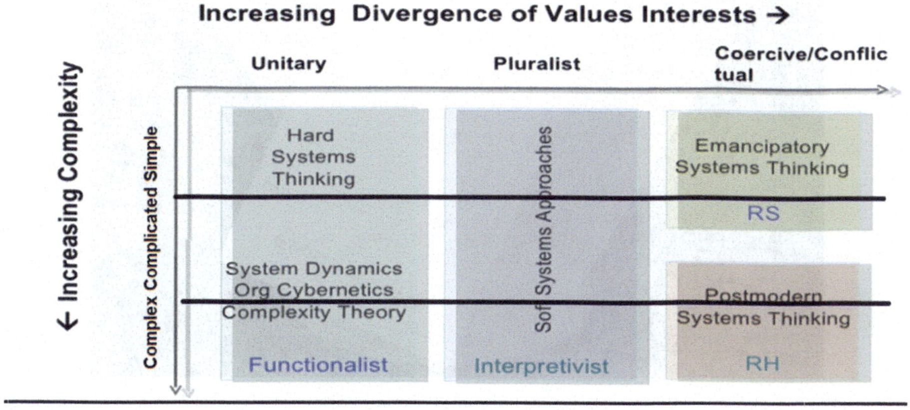

Fig. 2.2 Complexity divergence Michael Jackson (2019) critical systemic thinking and the management of complexity

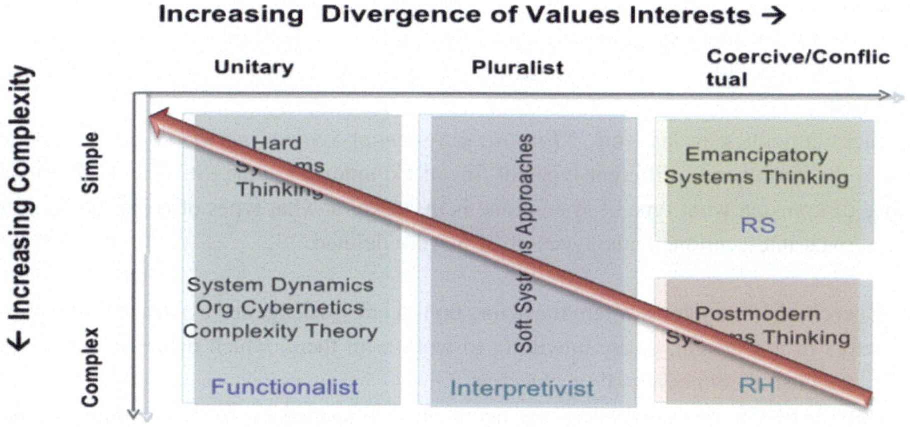

Fig. 2.3 Complexity divergence Michael Jackson (2019) critical systemic thinking and the management of complexity

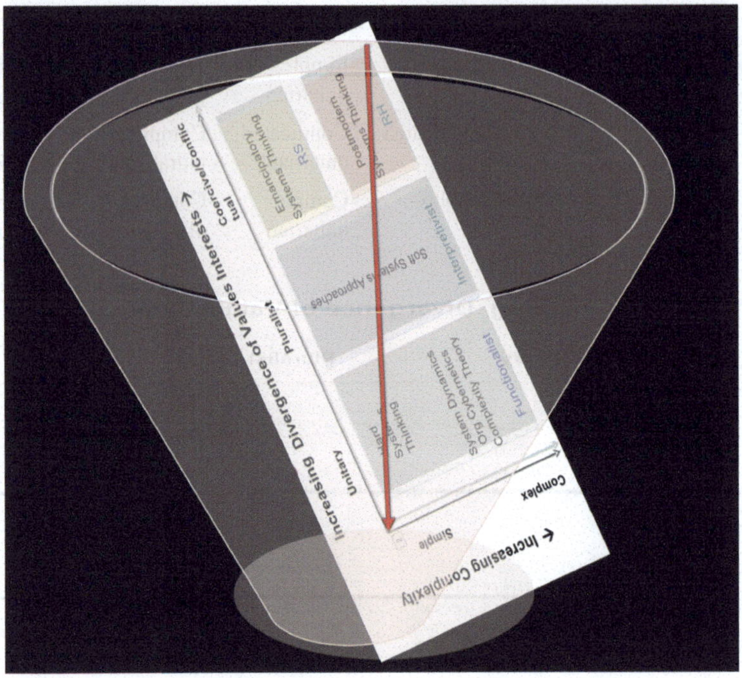

Fig. 2.4 Complexity funnel methodology proposal

application documents, and only identified one that described the mixing of traditional qualitative and quantitative methods. This indicates a possible gap in the current literature, and the Complexity Funnel methodology and its associated social transformation workshops seeks to contribute towards the generation of further empirical knowledge in this area.

Jackson defines, in his work, "*Towards a System of Systems methodologies*," (Jackson 1984) the existence of different types of Social Complex Systems (SCS) and the importance of knowing what type of system one is in, to know what types of tools can be used for a given intervention. Three types of SCS were defined:

- **Coercive SCS**: Systems where the actors do not acknowledge the existence of the other actors in the system, or are unwilling to work with them, which makes it impossible to establish a common goal or objective.
- **Pluralist SCS**: Systems where the actors have managed to set a common goal but disagree on how to reach that goal due to the existence of obstructions of both discrepancies and conflicts.
- **Unitarian SCS**: Systems where the actors have an objective and have managed to overcome the existing obstructions in the system to co-create a project that allows them

Fig. 2.5 Complexity funnel 2017 (Navarro and Dalrymple 2019)

to close the gaps between the current state of the system and the idealized design. By involving all the stakeholders, the intervention is legitimized, and is more likely to sustain.

Inspired by Habermas, Jackson, Ulrich, Midgley and Ackoff work on multi-methodologies and Systems Thinking, and knowing that the social complex systems (SCS)with which we work are not static systems but dynamic systems in permanent change where different perspectives coexist, both discrepancies and conflicts, we create a way of representing these systems recognizing these obstructions, identifying discrepancies and conflicts, mapping them and making them evident to all actors in the social system so that a common model of the system that could be legitimized by all actors was possible.

By managing to integrate into the general representation of the complex social system all the elements that could regenerate friction or conflicts between actors, we gradually managed to generate a common model of understanding, recognizing the points of inflection but allowing to unconsciously generate a space of joint work and of co-creation where all stakeholders can align to work for a common goal.

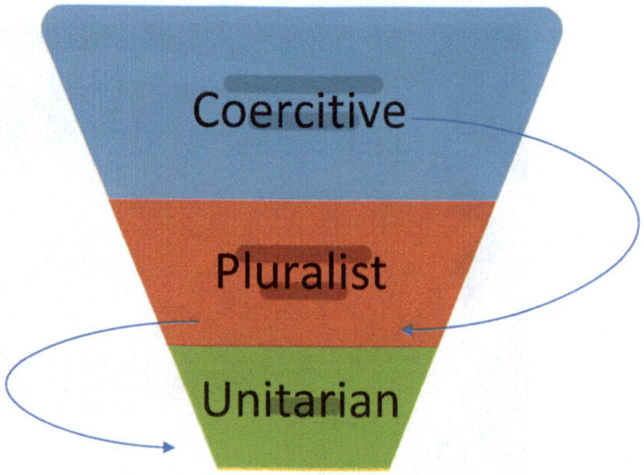

Fig. 2.6 Complexity funnell Navarro (2018) world engineering education forum Albuquerque New Mexico

The Complexity Funnel (CF) was created to act as a catalyst, accelerating the processes that reduce complexity in SCS, taking them from coercive to pluralistic and then to Unitarian (See Fig. 2.1). Decision makers are supported through this process, step by step using the social transformation workshops (STWs) (Fig. 2.6).

- **Coercive**: in this case, each actor identifies himself, but is not able to recognize the existence of the rest of them or they are simply not affected by teamwork, which makes it difficult to set goals and objectives for those who want to work.
- **Pluralist**: At this point, the objectives and goals have been identified and socialized collectively. However, we have not managed to reach an agreement on how to achieve them and present obstructions characterized by discrepancies and conflicts.
- **Unitarian**: Finally, the actors have already overcome the obstructions, which allows them to open the way to the creation of a project that allows them to reduce the gap between the social system and the idealized involved to each one of the interested parties (interested parties).

There is then a great opportunity to we engineer as we have the knowledge and practical skills available to improve the many forms of material injustice that are the root causes of the most violent conflicts on this planet.

In our particular case, critical systems thinking methods and methodologies are combined to help decision maker's deal with complex real world situations. Our purpose is both research and practical intervention. Real world situations are indeed complex and multi-dimensional, while particular research or intervention methodologies focus only on specific aspects the use of a combination of methodologies from different paradigms and methods together in a single intervention is used in the CFM. Research and intervention are not discrete events but more like a process that has phases, different types of activities that will predominate at different times. It encourages triangulation of research and generates more interesting and stimulating results.

Inspired by the number of successful reported interventions using Multi-methodologies, pluralism become not just more popular but necessary as part of the CFM.

The world we live in and we confront and seek to transform by our interventions as engineers is Multidimensional. Any real-world situation can be considered a complex interaction of substantively different elements.

There will be aspects that are relatively hard and observer independent, particularly material, and physical processes, that we can observe and model.

There will be aspects that are socially constituted, dependent on cultures, social practices, languages, and power structures, that we must come to share and participate in.

Finally, there will be aspects that are individual such as beliefs, values, fears, and emotions that we must try to express and understand. Meaning that different paradigms will focus on different aspects of a particular problem, highlighting some features and hiding others. Working with multiple paradigms looking through a variety of theoretical lenses employing different generic CST methodologies trying to capture as many different facets of a problem as possible and plan an intervention in a more informed and comprehensive manner as part of the CFM makes sense.

CFM is a process that is organized and proceed in a number of phases. Each phase involve different tasks and pose different problems for researchers and practitioners. This is why CFM has a wide spectrum of methodologies and methods available to face this problems situations using the STW´S.

Postmodern knowledge (Metanarratives) used in CFM refines our sensitivity to differences and reinforces our ability to tolerate the incommensurable.

The Lack of success of traditional methodologies and methods (hard systems approach) to deliver results while addressing complex problems made academics start questioning the old ways and made important advances in new approaches like the CFM giving birth to the pluralistic and multi-methodology approach we are looking for.

Multi-methodologies used in CFM allows a form of triangulation, were a decision maker can use different lines of sight to accurately fix a position. If a similar result is obtain by using different research approaches then researchers will have more confidence on the validity of their findings.in other words Looking a phenomenon from various perspectives provides a greater appreciation of its complexities. CFM proposal opposes the "isolationist" traditional approach. Accepting that different theoretical and methodological approaches are legitimate and can address problems situations differently is a first step towards pluralism. Recognizing the need to use different approaches in combination to tackle one problem situation because of its multidimensional characteristics that´s the real breakthrough to fully fledge pluralism.

Depending on the specifics of each intervention it might be important to alter the methods that were chosen during each phase of the CFM intervention. This will impose to decision makers to know or to be advice about how to combine approaches that may rest on different social theories. The complexity, heterogeneity and turbulence of many problem situations demand that pluralism must be exercised at three levels.

1. Methods
2. Methodologies
3. Paradigms.

Inspired by the TSI Total Systems Intervention CFM plan design and solve problem situations based on critical systems theory and the mix of methods from different generic systems thinking methodologies, giving practitioners an early guide on how to intervene coercive contexts and have a closer relation with social just like TSI, CFM is 100% committed with:

1. Complementarism
2. Human Well-being
3. Emancipation.

Just like TSI, CFM is a meta-methodology that employ in appropriate manner other systems methodologies it was develop as a proposal on how to put different systems methodologies to work in a coherent way according to their strengths and weaknesses, and also taking into account and integrating social conditions of the problem situation.

All this methods and methodologies were made available to decision makers in social complex systems so that their projects or the projects they were responsible for could embrace efficiency effectiveness and viability, looking forward to be able to promote mutual understanding among all stakeholders all actors in the systems and at the same time in parallel giving attention to issues of empowerment power relations and emancipation.

Following the 7 principles necessary for this Meta Methodology to work proposed by Flood and Jackson in 1991:

1. CFM promotes the existence of multiple perspectives and fundamental fixes. Problem situations are to complex and complicated to understand just from one perspective and the issues they throw up are too complex to be tackle by quick fixes. Quick fixes will fail.
2. CFM stimulates multiple perspective investigation: Problem situations and the concerns and issues they embody must be investigated from a wide variety of perspectives. This will allow decision maker to identify obstructions both discrepancies and conflicts.
3. CFM make a suitable choice of methodologies from 5 years of practical exercises both at a national and international level: Once the mayor issues are highlighted, a suitable choice of methodologies or methods is necessary to guide the intervention.

4. CFM highlights the strengths and weaknesses of each method and methodology to share this information with decision makers and use this knowledge together in combination with a social understanding of the main issues and concerns to guide the appropriate choice of methods and methodologies for the particular SCS that will be intervened.
5. In CFM different systems methods and Methodologies and different perspectives should be used in a complementary way to highlight and to be able to address different important aspects of the problems situations.
6. In CFM A systemic cycle of inquiry was set by TSI with a clear interaction back and forth between its three phases. Creativity Choice and Implementation.
7. In CFM All the actors, stakeholders, facilitators, and participants are involved and engaged at all the stages of the TSI process.

CFM as a pluralistic multi-methodologies approach look at problems situations as messes that cannot be understood from only one point of view but from a variety of perspectives. Obstructions will appeared both discrepancies and conflicts but once an agreement is reached between the actors about th existence of these obstructions they can start working together to solve it. There will be a common model shared by the actors were the obstructions will be highlighted, there will be agreements about the mayor issues and problems they are facing as part of the same system. The choice of methods and methodologies to address and manage this mess and tackling the main issues must be made in full knowledge of the strengths and weaknesses of the different systems approach.

Using Systems of Systems methodology SOSM decision makers were able to keep always in mind the idea of pluralism while choosing different methodologies or methods to be applied as part of the CFM. Different methodologies are used to address different aspects of the problem situations and to ensure that the technical practical and emancipatory interests are all given proper consideration. The nature of the mess can change during an intervention so it's important to keep under review and as an open dialogue the selection of methods and methodologies depending on how the implementation is going. A pluralistic multi-methodology approach as TSI and CFM seeks to guide decision makers in their interventions in a way that continuously address the major issues and problems faced in a complex social system, an organization or in a multi-agency situation.

Just as TSI meta-methodology CFM also has 3 phases,

1. Creativity,
2. Choice and
3. Implementation.

CFM Creativity phase seeks to highlight the mayor concerns and issues that surround the problem context. Different perspectives focus the attention on different aspects of the problem context. This ensures that decision makers gain a holistic appreciation of the problem and consider different perspectives that will draw their meaning from different paradigms. The key aspects of the problem situations are revealed and discussed. To do so we use the first two Social Transformation Workshops STW1 the web of Life and STW2 the rich picture.

CFM Choice phase, builds together in a participatory way a suitable intervention to transform the social system. Building an intervention strategy around a combination of systems methodologies. The choice depends on the unique characteristics of the problem situation that were highlighted in the creation phase and the particular strengths and weaknesses of the different Systems methodologies. The different methods and methodologies their strengths and weaknesses are explained to the decision makers so they can consciously choose a customized intervention.

CFM uses Jackson´s Systems of Systems Methodology SOSM that unearths the assumptions underlying different system approaches by asking what each assumes about the system in which it hopes to intervene and about the relationship between the stakeholders associated with that system. Doing this exercise, it is then possible to do an informed choice of a systems intervention strategy and the methods and methodologies to be used in that strategy. The most probable result is that there will be a more dominant methodology that will be supported by dependent methodologies to help with secondary problem areas for example.

CFM Implementation or execution phase. Tries to bring a positive change in the social complex system reducing social complexity and helping decision makers make better decisions. Been aware of the different possibilities offered by generic systems thinking methodologies and their methods and how they can complete each other in synergy, the problem situation may change over time, and it will be necessary to reassess the state of the problem situation by re-entering the creativity phase and questioning the original strategy or execution plan.

Just as proposed by Jackson CFM has dominant and dependent methodologies and switching between them helps with the practical difficulties associated with multi-paradigm working, problems that arise to be manage if an initial choice of dominant methodology is made, to run the intervention, with a dependent methodology or methodologies reflecting alternative paradigms in the background.

In the particular case of CFM, the dominant methodology is Interactive Planning and Idealized Design from Russell Ackoff (1981) The relation between dominant and dependent methodologies can then change as the intervention proceeds in order to maintain flexibility at the methodology level this allows the intervention to proceed in a theoretically informed way, making research possible, and protects paradigm diversity while being less confusing to participants. CFM is explicit about the initially dominant methodology "Interactive Planning or Idealized Design" but is also flexible, and ready to switch as necessary, other methods or methodologies are introduced based on different paradigms, as the situation demands

Society as seen by Althusser can be seen as a system that at various times is dominated by one of its main instances like economic or political or theoretical or ideological… among many others. In other to be able to change social totality decision makers should be able to understand the interrelations that exist between these main instances how they develop individually and how the dominance of one or another changes over time. Jackson replaces the notion of dominant instances presented by Althusser and replacing it with the idea of improving efficiency viability mutual understanding fairness among others. Which becomes the dominant concern in any problem situation for some ´period. With the right amount of time, they will eventually all be addressed but is just impossible to do it all at once.

Pluralistic multi-methodologies as CFM are iterative and systemic process it privileges those aspects that need to be addressed more urgently but it never stops is a permanent continuous improvement cycle in a complex and turbulent environment.

Problems are not static problems are dynamic and how problems are perceived by those actors that are part of the social complex system changes too, new intervention strategies need to be planned the only way to successively do it is to continually cycle around Creativity, Choice, and Implementation, changing methodologies that are dominant and dependent just as CFM does. This approach encourages highly creative thinking to really understand the unique characteristics of each social complex systems and the problem situations that exist within. Obstructions, discrepancies, and conflicts. In highly complex problem situations, it´s advisable to address at the same time different aspects by considering different perspectives. To do so decision makers can use CFM to vary between different systems methodologies in combination but having "Interactive Planning" IP or Idealized Design ID as dominant and the others (CSH, SSM, VSM., SD…) as supportive.

They are authors as Gerald Midgley that doesn't believe that TSI or CFM are meta-paradigmatic. Because they embody its own unique assumptions with which other theorists as him do not agree…related with Habermas theory of knowledge constitutive interest these theories rest on an untenable belief in the march of progress towards universal emancipation and risk ignoring the need for sustainable improvement that takes the natural environment into account. He also agrees with Ulrich that TSI has a restricted conception of the role of critique and only seemed to bring it into play when coercive contexts are identified… Another author Churchman states that critique is needed all time not least to identify cases of coercion… This is specially the case when working for a specific client when they are pressures to take organizational boundaries for granted and not to consider the impact of interventions on other stakeholders But boundaries drawn around a system determine what is taken into account, which stakeholders are considered, how improvement is defined and what actions are contemplated. boundaries of systems are not fixed by the structure of reality but by the knowledge and values of observers and so that in order to justify complex systems interventions we need to take in to account several different perspectives of a wide variety of stakeholders giving special attention to those that tend to be excluded or that that doesn't have a voice.

Ulrich took this idea from Churchman and the knowledge derived from Habermas work on discursive rationality and created Critical Systems Heuristic. CSH is used as a supportive methodology in CFM. His methodology offers a means of supporting rational professional practice that depends neither on the impossible sweeping in of all stakeholders nor the utopian vision of an ideal speech situation. Critical Systems Heuristic insist that appropriate boundary judgments can only be established through dialogue between those involved in designing the systems and those who are affected by the design. It also provides means to be able to detail the nature of different boundary judgments and allowing them to be rigorously debated.

Joining Churchman and Ulrich in their proposal but without losing those aspects of methodological pluralism Midgley propose Systems Intervention (SI) as a purposeful action by an agent to create change in relation to reflection upon boundaries.

It starts with a reflection upon boundary judgments because as Ulrich state boundary critique must not be subordinated to methodology choice but needs to precede it. The separation of the two methodological concerns of critique and emancipation is not tenable on critical grounds. Following this thoughts Midgley suggest starting by asking questions about boundary critique in any intervention because this way a new paradigm can be built around the ideas that appears with the boundary exercise. Which is very suited to CST interventions. Boundary critique enables decision makers to overcome subject-object dualism one of the most important enemy of systems thinking and one that has so far not been dealt with just yet. If the process of making boundary judgments is the most important part, rather than a particular kind of knowledge generating system, then subjects come to be defined in the same way as objects by a boundary judgment. Viewing Critical Systems interventions this way is unsustainable to say that the CFM is multi-paradigmatic but instead it can be said that the intervention has its own process paradigm based on boundary judgments. This paradigm will be subject to attack from non-believers especially those who continue to adhere to various forms of dualism and so CST is again subject to paradigm disputes. The exercise of process philosophy allow us to talk about the real world social construction and subjective understanding without contradicting ourselves it is preferable to realism, idealism and social constructions because it can offer an alternative grounding for most of the work from these other paradigms without the need to alter their most important contributions.

Then goes to a second part the "Creative Design Methods" were it makes use of the variety of available methods found in the systems literature and even beyond including methods from other areas to be able to create a flexible and responsible approach to intervention (Fig. 2.7).

The Complexity Funnel Methodology (CFM) is dissected in several Social Transformation Workshops (STW) that seek to reduce social complexity within Social Complex Systems (SCS) in terms of obstructions, both discrepancies and conflicts as defined by

Fig. 2.7 Complexity funnel methodology logo

Ackoff (Ackoff 2006) in order to co-create with the actors of these systems common models of understanding, which allows aligning interests of short, medium, and long term and give legitimacy to the planned interventions.

Among the different methods and methodologies used to apply the complexity funnel, 12 methods were established that we transformed into workshops to make available to decision makers in complex social systems.

2.2 Organizational Entropy Metric

We talked about the methodology proposal and how it should be applied, then we talked about how it is divided into several social transformation workshops (STW). However, very quickly the question arose as to how we could be sure that engineering interventions in these complex social systems were successful. Inspired by Ackoff work (Ackoff 1993) about obstructions both discrepancies and conflicts we define the success of our interventions in relation with the number of discrepancies and conflicts that could be resolved by the implementation of the different workshops.

Organizational Entropy Metric

We saw the different stages of ethics and the different elements that we must take into account to make a successful engineering intervention in complex social systems. However, we wanted to be able to evaluate the effectiveness of these interventions and the use of these workshops in terms of the social complexity of the systems measured in terms of obstructions, both conflicts and discrepancies as defined (Ackoff 1993). For this we proposed the metric of Organizational entropy that helps us to see how, as the social transformation workshops are implemented, both discrepancies are gradually resolved thanks to the secure dialogue between the different actors and conflicts thanks to the agreements between actors.

Entropy:

If we search the dictionary for the word Entropy we will find that.

"*Entropy is the second law of thermodynamics it states that systems tend to chaos over time*".

Nature, the Universe, our own existence is ruled by laws that are well known, laws that nothing and nobody can escape. This Laws are extremely precise, so precise they are expressed throw the language of mathematics. And Laws of nature we are yet to discover cannot be that different from the ones we already know. We are getting close to completing a description of the universe and understand how it works. Nevertheless, the scientific community had overlooked the relationship between those laws and complex social organizations behavior. In this paper a possible application of this concept of entropy will be used to explain the development of a new metric called "*Organizational Entropy*" to understand how engineering interventions can successfully reduce complexity within SCS.

In physics thermodynamics is the study of the movement of heat, specifically, the ways in which energy is transformed because as you know energy can take many forms. In case of SCS, energy is transferred in the form of resources, tasks, communications, responsibilities normally from the heads or decision makers to those responsible for the execution through a chain of command and communication.

In order to discuss social dynamics, we need to refer to a system. MIT professor Donnella Meadows refers to a system as;

> "*...a set of related components that work together in a particular environment to perform whatever functions are required to achieve the systems objective.*"
>
> (Meadows 2004, 2008).

The first thing we must know is that social dynamic systems does not tend toward equilibrium has thermodynamics systems do in physics. In physics equilibrium is where there is no more heat transfer, Social dynamic systems look forward to a state of equilibrium which means absence of obstructions, conflicts, or discrepancies but hey are continuously changing. This means that if we do not intervene these dynamic systems they will tend to disorder to failure and to disappear eventually. The job of decision makers as heads of this SCS is to make decisions that prevent their organizations to be disorganized to fail and to disappear, in this way they are in a permanent war against the second law of thermodynamics. The second law states that the sum of the entropy of a system and its environment must always grow. In other words, disorder always increases. And this seems strange but what it really means is that when there is concentration of energy, naturally the universe is responsible for dissipating it effectively.

Seeing it this way, decision makers in SCS are in a permanent war against the very laws that govern our universe. The macro and microenvironment of SCS is always changing, and this means that the heads of organizations must be constantly making decisions so that these organizations are able to adapt to these changes and conserve an equilibrium state. The success and failure of SCS depends then on the capacity and flexibility they must adapt to change.

Metric

It is very difficult to measure Complexity. If we search a dictionary for the definition of complexity, we will find something like.

> The behavior shown by a complex system.

Then if we search complex system, we will find something like.

> A system whose behavior exhibits complexity.

Complexity is difficult to define, it can mean different things to different people. A complex system can be a system which scientists believe to be complex. In this way Complexity Science can be seen as

> "*...the study of the phenomena which emerge from a collection of interactive objects.*"

In this sense even armed conflicts can be seen as a complex system were collective violent actions of different groups of people who are fighting for control of the same resources, land or political power creates complexity. Complexity science is supposed to help us solving wide range of important problems which face us as individuals and as a Society. Is a discipline in the making and that´s why we focus on possible application of it and where is heading rather than where it has been? Complexity Science is still being developed and its potential applications explored. Given this relatively immature nature of this emerging field I based this paper and my research in social complex systems (SCS) in my own academic research experience in Complexity, trying to give the readers a sense of the excitement of what is like to work at the forefront of this new discipline in such a challenging area of research and the challenges that comes with it.

We wanted to be able to calculate the success of our intervention, and by talking to the decision makers in the intervened systems they said that the success of the workshops rely on how they help reduce discrepancies and conflicts that exist within the systems.

We decide then to use obstructions as proxies that help us measure complexity in SCS.

We developed a new metric called "*Organizational Entropy*" (OE). Mapping these obstructions, discrepancies and conflicts that exist within the SSCs allows us to calculate the EO to give us an idea of the level of complexity of the SSC by studying how this complexity changes over time.

SCS are formed by elements those elements are both actors and variables. Actors make decisions based on their perceptions of the system, and those decisions affects the system´s variables and these variables change the system itself. Systems are dynamic and adaptative. At the head of these systems, they are decision maker's actors that oversee anticipating possible changes that may occur both inside and outside their systems and that may affect their functioning. They are responsible for the success or failure of a System. They are responsible for making decisions that affect many actors and key variables based on incomplete information and limited resources. It is not an easy job. This decision makers oversee avoiding that chaos an keeping the systems working the best they can. To do so they must be able to identify obstruction that may exist in the system. Ackoff (1981) defines two different types of obstructions. Discrepancies and Conflicts.

2.2.1 Obstructions

Obstructions are characteristics and properties of the SCS that obstruct its progress or resist change they are classified in two categories, Discrepancies and Conflicts. This obstruction prevents the organization from making adaptive changes to the always changing conditions. And if they are not address on time the SCS could eventually destroy itself if it were to continue doing what it is doing currently.

2.2.1.1 Discrepancies

When different actors have a different idea about what´s happening in the system the causes and consequences. About what things really are and what they believe they are.

They are different types of Discrepancies, 5 to be exact:

Discrepancies Type 1 the aims of the SCS actors:

An objective observer can detect that a SCS actor does not always address the ends, goals, objectives and ideals it claims to pursue. Often there is a big difference between what is preached and what is practiced.

Discrepancies Type 2. The means of the SCS actors:

The second type of inconsistency or discrepancy has to do with the means that a SCS actor proclaims or believes it uses in the pursuit of its purposes, and the means it uses. There is a big difference between what the staff really does and what they think they are doing or what they are supposed to do, that is why they say what: either assumptions or beliefs are changed or what is thought to be done. It must be done.

Discrepancies Type 3. Organization of resources:

The third type of discrepancy is related to beliefs about the quality, quantity, and use of the resources available to the SCS actors. These resources are staff, facilities, and equipment; materials and energy; Informality and money.

Discrepancies Type 4. Organizational structure and administration:

The fourth type of discrepancy is related to the way in which the pursuit of the objectives is organized and managed.

Discrepancies Type 5. The participants of the System and the environment:

The last type of discrepancy is related to the assumptions about the participants and the environment. Frequently such assumptions serve as the basis for policies and strategies.

2.2.1.2 Conflicts

Conflicts exist when two or more desires interact in such a way that progress towards one of them prevents progress towards others.

Midgley states that conflict between groups arise when they hold different views about the facts and values that are relevant to the situation they find themselves in (Midgley 2011)

They are different types of Conflicts, 7 to be exact.

The different types of conflicts that take place during the development of a company are detailed below:

Conflict type 1 internal conflicts of individuals:

There is increasing awareness in SCS of the harmful effects of internal conflicts of actors at all levels. A growing number of evidence clearly demonstrates that alcoholism, and drug addiction, as well as family, financial and legal problems among many others reduce the ability of individuals to perform efficiently.

Conflict type 2: Conflicts between individuals:

These conflicts may be caused by differences in personality and disagreement regarding the objectives and the means to achieve them. Whatever its origin, conflicts of this kind can reduce collective and individual efficiency, as well as the ability to improve it.

Conflict type 3. Conflicts between individuals and the SCS or parts of it:

When such conflicts arise, people often try to obstruct, consciously or unconsciously, the pursuit of the SCS objectives, and often achieve it. Actors who have this tendency are called alienated. Disposal is a matter of mood within the team and tends to be contagious.

Conflict type 4. Conflicts within the units:

The high authorities or the decision makers impose mutually conflicting objective units on their units. For example, ask to some actors to advertise a product to achieve profits for the community and on the other hand also ask the community to accelerate its pace of work so that the system can achieve its objectives.

Conflict type 5. Conflicts between deferent decision makers:

This type of conflict occurs with frequency when different decision makers of the same level are assigned objectives that conflict with the objectives of other decision makers of the same level.

Conflict type 6. Conflicts within the SCS as a whole:

This type of conflict can be described with the following example: Many SCS simultaneously seek to maintain good relations with all the actors and, however depending on limited resources, make periodic staff cuts or stop paying some of them. Other SCS want to increase their market but are not willing to increase the number of actors helping with marketing management. It can be mentioned that having conflicting goals is not uncommon; however, if the attempt to solve them fails, they can obstruct the development of the SCS itself.

Conflict type 7. Conflicts between the SCS and external groups:

These conflicts take place particularly with civic groups and with government agencies and are often used as an excuse to act inefficiently. However, inefficiency even if you apologize, cannot contribute to the development of the company.

To face these obstructions the person in charge the decision maker must create "Safe Dialogue Spaces" SDS to address the discrepancies and conflicts so that the other actors acknowledge different point of views about the same topics and create agreements to face conflicts. When we identify a conflict the most important thing is to understand that none of the actors involved in conflicts is going to be 100% happy about the final outcome but we need to find balance between the different points of view the different perceptions and different paradigms.

To solve discrepancies is easier than to solve conflicts, since to resolve a discrepancy between actors, it is enough to generate a safe space for dialogue where the actors feel comfortable sharing their understanding of the system and entering into a constructive dialogue with other actors. The conflict is more complicated since it requires the creation of an agreement that in no case will benefit 100% of any of the authors.

Our investigation of the cases of application of the EC methodology and the TTS workshops suggests that it takes almost 3 times more time and effort to resolve a conflict than a discrepancy, so the Organizational Entropy (EO) Navarro 2018) is calculated as:

$$\Delta SOi = \sum_{i=1}^{n} n * Di + 3n * \sum_{i=1}^{n} Ci$$

Where ΔSOi is the variation of the EO metric, n is the number of system connections (Interrelations, D is the number of mapped discrepancies, and C is the number of mapped conflicts. This new metric was presented at the World Engineering Education Forum (WEEF) in the city of Albuquerque in the United States in 2018 whose main theme was Peace Engineering whose organizer was (Jordan 2018). Organizational entropy then allows us to map the impact of TTS workshops on system complexity and is a good proxy for how engineering methodologies and concepts are helping to reduce the complexity of the systems studied.

Entropy is a function of state of a system it gives us an idea of the disorder that exist within the system, the higher the entropy, the more the system is disordered. Entropy measures the spontaneous dispersal of energy in other words how much energy is spread out in a process, or how widely spread out it becomes at a specific temperature. The second law states that in any process that involves a cycle like the dynamic processes the entropy of the system will either stay the same or increase. Organizational Entropy in the other hand can be reduced.

The Complexity or disorder existing in SCS measure as the number of obstructions can be decreased depending on the decision maker´s actions. With less complexity it´s easier for decision makers in SCS to make difficult decisions because they will have more information and more understanding about the effects of those decisions.

In physics Entropy is represented by the letter S and the units are KJ/mol.K. Organizational entropy is represented by the letters ΔOS and the units are Obs (# of obstructions).

This obs correspond to the number of obstructions both discrepancies and conflicts that we identify between the different actors in the studied SCS.

In Physics ΔS give us an idea of the disorder degree of a system. For example, in the physic reaction of water condensation

$$H_2 + O_2 => H_2O$$

the entropy level is reduced because we pass from a gas to.

In SCS the entropy is reduced if we manage to solve some of the discrepancies throw the dialogue and some of the conflict throw the creation of agreements between the affected actors. They are two ways in which we can reduce complexity.

When different actors have different interest on the behavior of a variable in the system.

Through the use of the Complexity Funnel Methodology CFM and the Social Transformation Workshops STW it is possible to reduce the Organizational Entropy that exists within Complex Social Systems CSS helping decision makers who have the difficult task of making decisions that will affect several actors based on incomplete information and limited resources.

References

Acero, A (2021) A Methodology to Foster Socially Just Practices in Engineering: A Systemic Approach—Universidad de los Andes

Ackoff RL. (1981). Creating the Corporate Future: Plan or be planned for. John Wiley: New York.

Ackoff, R. (1993). Idealised Design: Creative Corporate Visioning. Omega: The InternationalJournal of Management Science, 21, 401–410.

Althusser L & Balivar E (1970) Reading Capital London New left books.

Boyd A, Brown M, Midgley G (1999) Home and Away:Developing Services with Young People.

Chowdhury R & Jangle N (2018) Critical Systems Thinking towards enhancing community engagement in micro insurance Global Journal of flexible systems Management 19: 209–224.

Churchman C.W (1979a) The Systems Approach 2e New York : Dell Publishing

Churchman C.W (1979b) The Systems Approach and its Enemies. New York: Basic Books.

Churchman CW (1979c) Paradise regained. Ahope for the future of systems design education .In:Education in Systems Science (ed B.A. Bayraktar et al.) 17–22 London: Taylor & Francis.

Cohen,C and Midgley G (1994) The North Humberside Diversion from Custody Project for Mentally Disordered Offenders. Hull:Center for Systems Studies University of Hull.

Eden, C., Ackerman, F., Bryson, J.M (2009) Integrating modes of policy analysis and strategic management practice:requisite elements and dilemmas. Journal of the operational research Society# 60 p 2–13.

Flood R.L & Jackson M (1991) Crtical Systems thinking: Directed Readings Chichester Wiley

Henao F & Franco LA (2016) Unpacking multimethodology impacts of a community development intervention. European Journal of Operation Research 253 681–696.

Howick, S., & Ackerman, F. (2011). Mixing OR methods in practice: Past, present and future directions. European Journal of Operational Research, 215, 503–511.

Jackson, M.C. (2000). Systems Approaches to Management. Kluwer/Plenum, New York.

Jackson, MC (2001) Critical Systems thinking and practice. European Journal of Operational Research N 128 P 233–244

Jackson, M.C and Keys, P (1984) towards a system of systems methodologies. Journal of the operational research society 35:473-486.

Jackson, M.C. (1987) Present positions and future prospects in management science. Omega 455–466.

Jackson MC. (2003). Systems Thinking: Creative Holism for Managers.

Jackson M (2019) Critical Systems Thinking and the Management of Complexity. Responsible leadership for a complex world.

Jordan R (2018) Peace Engineering – The Concept of Sustained Growth https://www.istec.org/loslibertadores/wp-content/uploads/sites/7/2018/09/Peace-Engineering-%E2%80%93-The-Concept-of-Sustained-Growth.pdf.

Lyotard JF (1984) The post modern condition :A reort on knowledge Manchester University Press.

Meadows DH (2004) Limits of Growth : The 30-year update.

Meadows DH. (2008). Thinking in Systems – A primer. Earthscan: London.

Midgley, G. (1996a). What is this thing called critical systems thinking? In R.L .Flood, & N.R.A. Romm (Eds.), Critical Systems Thinking: Current Research and Practice New York: Plenum.

Midgley, G. (1996b). The ideal of unity and the practice of pluralism in systems science. In, Critical Systems Thinking: Current Research and Practice. Flood, R.L. and Romm, N.R.A. (Eds.). Plenum, New York.

Midgley G. (2000). Systemic Intervention: Philosophy, Methodology, and Practice. Kluwer/Plenum: New York.

Midgley, G. (2014) Systemic Intervention, Research Memorandum 95. Hull: Center for Systems Studies, Hull University Business School.

Mingers, J. (1997a) Multi-paradigm multimethodology. In MultiMethodology (ed Mingers J and Gill A) P 1–17 Chichester: Wiley.

Mingers, J. (1997b) Towards critical pluralism. In MultiMethodology, p 407–440. Chichester: Wiley.

Mingers, J. (2014). Systems Thinking, critical realism and Philosophy, Kindle Edition. London: Routledge.

Mingers, J. & Brocklesby (1997) Multimethodology :towards a framework for mixing methodologies. Omega 25: 489-509.

Mingers, J. & Gill, A (1997) Multimetodology: the Theory and practice of combining Management Science Methodologies. Chichester Wiley

Mingers. J. & White, L.(2010) A review of the recent contribution of systems thinking to operational research and management science . European Journal of Operational Research 207:1147-1161.

Munro I & Mingers J (2002) The use of multimethodology in practice –results from a survey of practitioners Journal of Operational Research Society 53: 369–378.

Navarro, CA and Dalrymple, O (2019) Co-creation of a systemic model to support community engagement projects

Rosenhead, J. (1986). Custom and practice. Journal of the Operational Research Society, 37, 335–343.

Ulrich, W (2003) Beyond Methodology Choice: Critical Systems Thinking as critically systemic discourse. Journal of Operational Research Society N 54 p 325–342

Zlatanovic D (2017) A multi-methodological approach to complex problem solving:The case of Serbian Enterprise. System 5.

3 Social Transformation Workshops

3.1 Social Transformation Workshop STW#1 "The Web of Life"

STW#1: The Web of life

About this workshop

At its simplest, The Web of Life Workshop allows a group to observe how the parts of a system of interest are interconnected. As they participate in this exercise, they see that the system at hand, whether it be the physical climate system, or the behavioral system related to a specific climate change intervention or policy, is typically not made up of straight lines of causality. Instead, it is made up of patterns of connection and interaction that better resemble loops, webs, and networks. In most situations, we cannot see these interconnections, we have to imagine them. The Web of Life workshop gives participants an opportunity to begin to appreciate and "make visible" the often-complex patterns of interconnection that comprise some of the more perplexing challenges associated with the problems we are trying to address as a group. In addition, this exercise can help intact teams "see" and better understand the interdependencies and the connections that exist among themselves and within the larger system (such as their organization and their community) (Photo 3.1).

To Run this Workshop:

Number of people: 8 people works well. You may want to experiment with larger numbers.

If you use larger numbers, make visible signs with the names and/or pictures of the variable to put around people's necks, like name tags.

Time: Approximately 15 to 30 min, depending on the number of people and the length of the debrief.

C. A. Forero Navarro, *Untangling Complexity—Peace Building Engineering*,
Synthesis Lectures on Engineers, Technology, & Society 29,
https://doi.org/10.1007/978-3-031-37038-0_3

Photo 3.1 Taller STW#1 the web of life Tijuana Mexico 2019

Space: Enough space so your group can stand, nearly shoulder-to-shoulder, in a circle.

Equipment: Long ball of yarn (or string or surveyor's tape), of different colors. Make sure it will unravel easily. A flip chart, index cards, sticky tape, and colored markers.

Set up: Organize the group into a circle. Give the facilitator the large ball of colorful yarn.

Before meeting all the actors, we should ask them to reflect on the actors and variables that they consider key to the system of which they are all part. At the beginning of the workshop, attendees are asked in pairs to socialize the list of actors and variables with their partner and add to their list the actors that they had not previously identified and think that it was because they had not been able to identify these actors.

Each actor can choose to represent himself or to represent one of the actors identified in the common model. The web of life network begins to be woven from the relationships that appear between these actors and the key variables of the system that appear as a product of the dialogue facilitated by the facilitator. The discourse will lead the actors not only to identify actors but also important variables according to which these actors make decisions that will affect the system.

Once all the attendees have an actor or a variable and this is interrelated by means of causal links with the wool thread, we put the "web or f life" network on the floor. for the first time the participants will then be able to see a common model of the social system to which everyone belongs and how everyone is interrelated.

We make a list of the variables identified in the system:

Variables:

Next, we choose together a variable that is the Variable of Interest of the system, The Main Variable and this variable should allow us to say whether the system is improving or not as a whole. The make them think about the possible main variable we can ask them which variable we should look at 5 years from now to see if our intervention was successful.

Variable of interest:

In a second part of this workshop with decision makers we can do the exercise of thinking about what can be a possible metric for these variables or a possible proxy variable.

Variable	Metric

Para el sistema social del cual usted hace parte se le pide que haga un primer ejercicio de identificaci6n de actores relevantes.

After the identification of the main variable, we immediately ask each actor to think about the role and interest that each actor has, those actors that represented a variable can team up with actor of their preference.

We gave each group a sheet of paper with tree columns and ask them to reflect on the role and interest of each actor. We can give them 20–30 min to do so.

Actor	Rol	Interest

After the 20–30 min we will ask each group to quickly shared what that they believe to be the roles and interests of the actor they represent and the others will complement that answer in a dialogue space directed by the facilitator. In a second part with the decision makers we can differentiate these actors between: Decision makers, Experts, Executors, Affected (Beneficiaries or Affected) and others.

Roles	Actors
Tomadores de decisión	
Expertos	
Ejecutores	
Afectados Beneficiario	
Afectados Victimas	

The following questions must be answered individually
What do you think is the problem?
Why is a problem?
Who is a Problem for?
Why is it worth solving?

3.2 Social Transformation Workshop STW#2 "Rich Picture"

STW#2 Rich Picture

Qualitative Pictures:

What is really interesting is that some of these methods primarily doesn´t involves words. Sometimes (Photo 3.2).

Photo 3.2 5 Live University of San Diego California United States 2019 rich picture

"A picture is worth a thousand words".

The Rich Picture methodology is a visual and participatory approach used to understand and represent complex situations or problems. It was developed by Peter Checkland as part of Soft Systems Methodology (SSM), a problem-solving and organizational learning approach. The Rich Picture methodology involves creating a detailed and holistic visual representation of a problem or situation. It allows stakeholders to collaboratively depict their perceptions, concerns, relationships, and dynamics related to the issue at hand. The resulting visual representation is often a hand-drawn or digitally created picture that captures the multiple perspectives and complexities of the situation.

In any system, how can we make sense of the whole system and of the role of each element in the system?

One way is to make a picture, or a pictorical representation of it.

Nearly a hundred years ago, in order to understand how organisms evolve over a rough fitness landscape, Sewall Wright (1986) proposed picturing the path of their evolution over a series of steps, boxes for the steps, arrows for the paths through them. To make evolutionary sense, he established rules on how to trace a path along the arrows (Acyclic graphs).

System of Systems Methodology

A Simple System will be perceived to consist of a small number of elements, and the interactions between these elements will be few, or at least regular.

A Complex System will, on the other hand, be seen as being composed of many elements, and these will be highly interrelated.

A Complicated System will be seen as being composed of many elements, and these will be highly interrelated but cannot really establish how.

> "*Soft Systems Methodology SSM is an approach for tacking problematical, messy situations of all kinds. It is an action oriented process of inquiry into problematic situations in which users learn their way from finding out about the situation to taking action to improve it. The learning emerges via an organized process in which the situation is explored using a set of models of purposeful action as intellectual devices, or tools to inform and structure discussion about a situation and how it might be improved.*"
>
> (Checkland and Poulter 2010a, b).

> "*...The world as a set of systems, and intervention in the world as a process of designing or redesigning systems.*"
>
> (Checkland and Poulter 2006).

Soft systems thinkers offer the notion of "system" to all stakeholders of an organization as a construct that they can employ to enhance their creativity and improve their chances of achieving what they want through their organizations. For authors like Churchman for example, systems are seeing as a concept that could be used to increase human purposefulness by helping facilitate the learning process through dialectical method.

Checkland sees his soft systems approach as viewing social reality as the result of a never-ending process of iteration between actors with multiple values and interests.

In SSM we move from relevant systems to root definitions to conceptual models. The participants are encouraged to build their own models from their own worldviews and began to appreciate the role these could play in enabling a better, more focused debate to take place. This important individual previous work will make them more willing to participate and engage in meaningful conversations.

SSM is a mature methodology capable of intervening in soft unstructured problematic situations were answering questions about what we should do is a significant as determining how to do it. It has made Systems thinking more practical and not just theoretical which made it more relevant to decision makers and managers in charge of complex systems.

> "*SSM has reoriented an entire discipline and touched the lives of thousands of people... soft thinking is now completely taken for granted within Systems discipline.*"
>
> (Mingers-Ramage and Shipp 2009)

Academic and practitioners have embraced SSM as a very practical and successful approach to messy and wicked problems and has spread its influence worldwide Ledington and Donaldson (1997); Hanafizadeh and Mehrabioun (2018); Van de Water et al. (2007); Mingers and Tylor (1992) show different successful applications of this methodology worldwide and its relevance in different fields.

About this workshop
At its simplest, The Rich Picture allows us to have a drawing image of the common model built by the community where all the elements, both actors and variables, are intertwined and this allow several features of the perceived problematic situation to be set down pictorially for everyone to see. Artist often uses drawings to provide arbitrary spurs to invention. The process of drawing is particularly conducive to associative thinking.

> *"Rich pictures can be great they allow groups to explore their subconscious, their occult sentiments and conflicted understandings."*
>
> Bell and Morse (2013).

To Run this Workshop:
Number of people: The rich picture can be created individually or in groups to be presented to the other actors.

Time: We can give them a day a week or a month depending on the availability of the actors and the time of our next meeting.

Space: We need a safe dialogue space were each actor or group of actors can safely share their Rich Picture.

Equipment: Each actor can choose the best way to share their rich picture.

Set up: Each actor can choose at their discretion the best way to share their rich picture can be on a large sheet of paper with magazine clippings or drawings or it can be in a power point, there are no limits to creativity, what is really important is the lecture in which they identify the relationships between actors and variables.

Suggested Rules:

Don't Structure your rich picture in any way.

Don't use too many words let the drawing speak by itself.

Don't exclude what you consider to be any relevant observations about culture emotions and values.

Include a representation of yourself because you are part of the messy situation.

Include a title & Date.

3.3 Social Transformation Workshop STW#3 "Critical Systems Heuristic"

STW#3 Critical Systems Heuristic

Critical Systems Heuristic Methodology:

"Critical Systems Heuristic" refers to a method or approach that combines critical systems thinking with a heuristic approach. Heuristics are mental shortcuts or practical rules of thumb used to simplify complex problem-solving processes. In the context of critical systems thinking, which emphasizes questioning assumptions, power dynamics, and ethical considerations within systems, a "Critical Systems Heuristic" could be an approach that incorporates heuristics to guide the analysis and understanding of complex systems.

According to Ulrich, the main principle of his methodology is the boundary critique principle (Figs. 3.1 and 3.2).

> "*...a claim meaning and how valid it is depending on its reference system, the boundary judgments that informs its view of relevant facts and values and thus its empirical and normative selectivity.*"
>
> (Ulrich 2012)

The boundary critique face 5 tasks:

1. Identify sources of selectivity in a system by developing the boundary judgments.
2. Asking about the practical and ethical implications of the boundary judgments.
3. Considering alternative answers to the boundary questions.
4. Looking for mutual understanding between different actors of the system, increasing tolerance and creating a shared perspective.
5. Challenging the claims of different actors that try to impose their boundary judgments to others.

"Reference Systems" as Ulrich calls them referring to their own view of a situation in the real world refers to the Weltanschauung define by Churchman or the Appreciative System from Vickers or Boulding's "image" of a system. They are selective in what they decide to highlight from a specific system leading unwillingly to specific claims on possible interventions to solve the problematic situations (Fig. 3.3).

The critical systems methodology think through the relationship between this 3 elements in what is called an eternal triangle, a continuous improving cycle, The relation between these elements and its analysis is called Systemic Triangulation (Jackson 2019) These reflection can guide the actors to important positive changes in the reference systems.

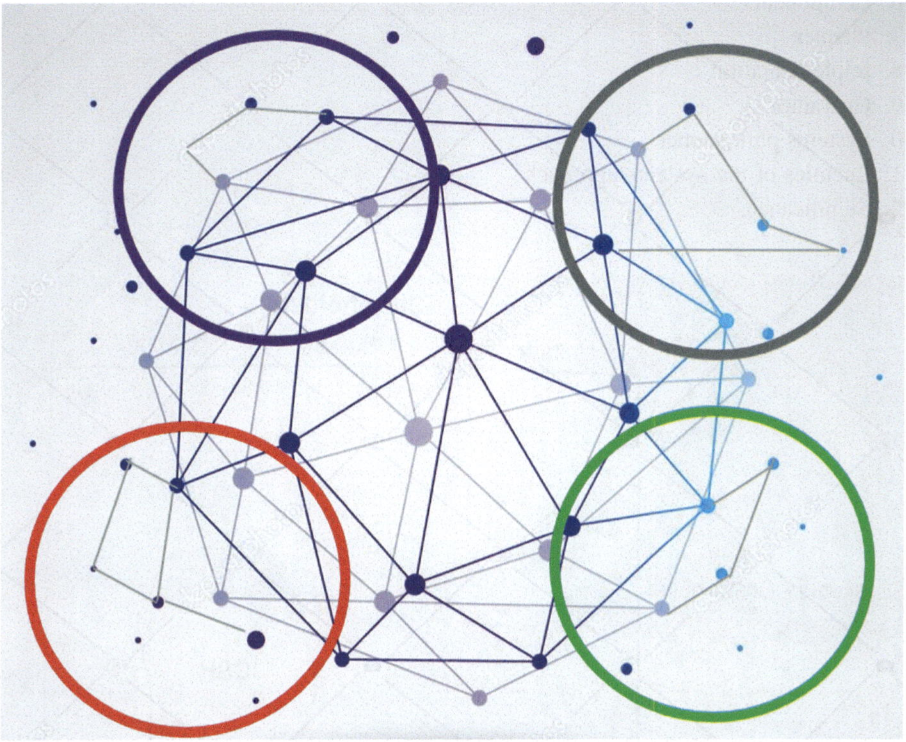

Fig. 3.1 Critical systems heuristic 2017 Colombia

Ulrich explains that social system designers inevitably come up against the basic forms of human intentionality among which we have self-determination, self-consciousness and self-reflexibility (Ulrich 1983).

Giving this Ulrich states that we need to add an important dimension when working to improve social complex systems, the dimension of "purposefulness" and design or redesign the intervened reference systems to have a purpose and become purposeful systems.

Churchman itself inspired by Kant propose 9 necessary conditions for a system to be purposeful and then in later work Churchman extended that same list to 12 conditions, the last 3 conditions being related to the lack of comprehensiveness of the systems approach.

1. Client
2. Purpose
3. Measure of performance
4. Decision Maker
5. Environment

6. Components
7. Planner
8. Implementation
9. Guarantor
10. Systems philosopher
11. Enemies of the systems approach
12. Significance.

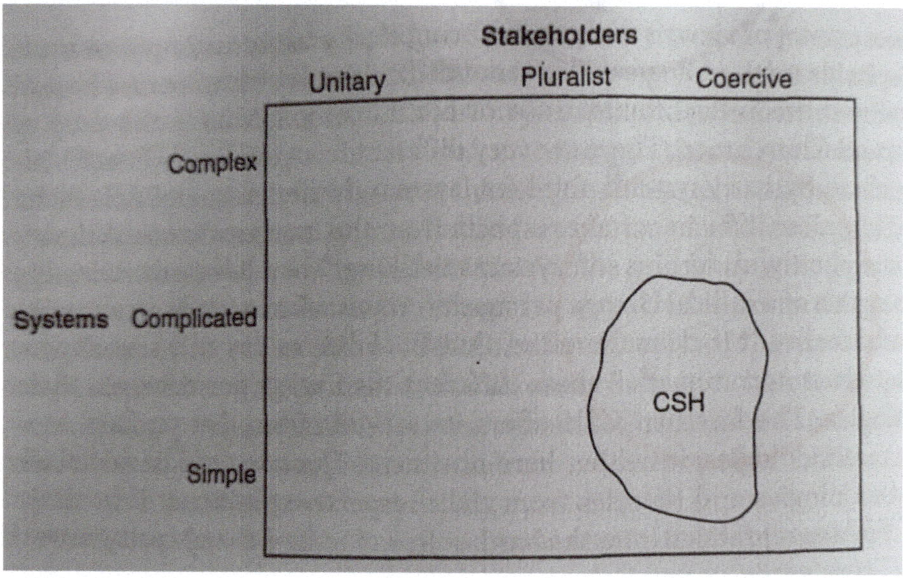

Fig. 3.2 Position of critical systems heuristic on the systems of systems methodology

Fig. 3.3 Eternal triangle–systemic triangulation. Own elaboration Navarro CA 2021

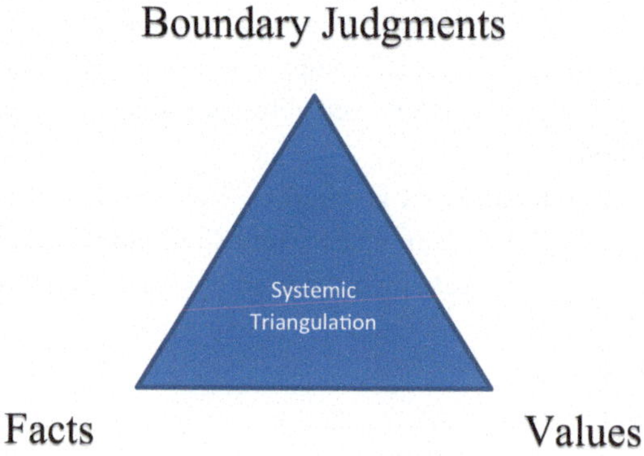

> *"Churchman idea was to employ the categories for tracing the different interpretations and valuations to which one and the same set of data about the problem situation lends itself, depending on the observer's world views and needs."*
>
> (Ulrich 1988a, b)

Inspired by Churchman Ulrich presents his own 12 boundary categories he replaced the last 3 categories

10. Systems philosopher
11. Enemies of the systems approach
12. Significance.

With

10. Witness
11. Emancipation
12. Worldview.

He decided to change these 3 categories because he believed the new ones could be used to reveal the selectivity of designs rather than just point to the lack of comprehensiveness of the systems approach as a whole. This way we will be critically reflecting on the reference system.

Ulrich's Categories are showed in the next figure.

The different sources of influence of a reference system are represented. The four groups, sources of motivation, of power expertise and Legitimacy bring to bear in a purposeful system. The 12 original Ulrich's categories lead to 12 boundary questions 3 four each group. The 12 questions are asked in both the is mode and should be mode how things are reality and how things should be in an idealized reference system.

The primary method developed to serve the Critical Systems Heuristic Methodology, even if the questions are so interlinked that it is possible to start with any of them but Ulrich and Reynolds (2010) have set a standard sequence for the 12 questions.

It also suggests is better to start the questions with the "ought" the ideal even if it is utopic. Because to think about the future the hopes and dreams is motivational for them.

During a review of more than 11 publications since 1985 to 1996 Midgley states that in his opinion.

> *"The successful use of CSH depends on a situation already existing in which participative debate is possible.... CSH can be use only when communication is possible, either face to face via an arbitrator, and only then if all participants are able to handle involvement in rational argumentation."*
>
> (Midgley 1997).

Midgley argues that coercion is usually characterized by closure of debate, where those in the head the decision makers those with power refuse to talk to others (Midgley 2000).

> *"CSH, which depends upon the possibility of communication between stake holders groups becomes redundant when coercion is experienced."*
>
> *Until the publication of Ulrich CSH there was no methodology or systems approach that provided the necessary means for a critical reflection upon the pursued goals. CSH possesses a unique ability to bring disadvantaged stakeholders into the debate providing them with the opportunity to state their case with the same confidence and competence as decision makers. This is way this methodology sill a huge gap in systems thinking. This is why this is one of the most important methodologies when we are addressing "coercive complexity".*
>
> *"… once the selectivity of claims has become transparent, democratically institutionalized processes of decision-making can work in a meaningful way."*

(Ulrich 2005).

Midgley states that Critical Systems Thinking could do more to help if it added to its already extensive repertoire means of supporting campaigning and non-violent direct action (Midgley 1997) Checkland also states that in coercive situations, no mere methodology which is only words on paper can guarantee the overthrow of a tyrant (Checkland 2003).

About this workshop

At its simplest, The Critical systems heuristic workshop understands that they are many perspectives on any problem. We all conceptualize problems as something that we feel we can do something about and suggest the type of solution that pleases us most. But synthesis is possible between different positions. This workshop seeks to subsume all different actors' perceptions in one worldview or as Germans calls it a "Welstanschauung". The systems approach enables every actor to appreciate as fully as possible his own view of the Social complex Systems they are a part of. By listening seriously to others views and questioning their own mental model about the system and the common model that was created with the other actors.

To Run this Workshop:

Number of people: It can be done individually, in groups or just by the decision makers and then presented to other actors.

Time: 1 day a week or a month depending on the date of the new meeting.

Space: We need a safe dialogue space were each actor or group of actors can safely share their Rich Picture.

Equipment: Each actor can choose at their discretion the best way to share their answers to the critical systems heuristic 24 questions.

Set up: We need to set a safe space so that all actors can share their answer to the question, we will feed the common model of the SCS with the discussion that follows. ***ace***

Define or list the characteristics of the studied system model is: (The characteristics of the real system)____

____

____

Answering individually the 12 questions in the model is.

The model is representing reality and answers the 3 questions for each of the 4 border limits.

I. Motivation Limit

Question 1: Who are the customers benefited or affected by the S system?

Question 2: What is the purpose of the S system? Measured in terms of the consequences of the same (Variable of interest) and not the intentions

Question 3: How is the success of the system measured?

II. Limit of Power

Question 4: who is or who are the decision makers?

Question 5: what decisions does the decision maker control (variables)?

Question 6: what decisions does the decision maker not control (Variables)?

III. Limit of Expertise

Question 7: who is an expert in the field?

Question 8: what kind of expertise do you have?

Question 9: how it is ensured that the expert is heard by the decision maker

IV. Legitimacy Limit

Question 10: who represents the interests of those affected?

Question 11: how are the interests of those affected taken into account? (Emancipation of experts and decision makers)

Question 12: what is the dominant worldview in the design of the S system (Whose worldview?

The answer to these possible questions can make the common model improve with the appearance of new actors or new variables that have not been taken into account.

Name the new actors or the new variables that appeared when answering the 12 questions of the model.

2. Model Must Be Now the same 12 questions will be answered, but from the ideal model. Ackoff's Idealized Design:

Define the characteristics of the system in the ideal state. (Without any restictions)

Answering the 3 questions for each of the 4 boundary limits for the model should be

I. Motivation Limit

Question 1: Who should be the customers benefited or affected by the S system?

Question 2: What should be the purpose of system S? Measured in terms of the consequences of the same (Variable of interest) and not the intentions

Question 3: How is the success of the systems measured?

II. Limit of Power

Question 4: who should be the decision makers?

Question 5: what decisions should the decision maker control (variables)?

Question 6: what decisions should the decision maker not control (Variables)?

III. Limit of Expertise

Question 7: who should be an expert in the field?

Question 8: what kind of expertise should I have?

Question 9: how should it be ensured that the expert is heard by the decision maker?

IV. Legitimacy Limit

Question 10: who should represent the interests of those affected?

Question 11: how are the interests of those affected taken into account? (Emancipation of experts and decision makers)

Question 12: what should be the dominant worldview in the design of the S system (Whose worldview?

That new actors or Variables appeared within the system.

Add the new Actors and variables to the common rich box.

New Common model.

3.4 Social Transformation Workshop STW#4 "Social Cartography"

STW#4 Social Cartography-Cartografía Social

Social cartography, also known as participatory mapping or community mapping, is a methodology that involves the collaborative creation of maps or visual representations to capture and communicate the knowledge, experiences, and perspectives of a community or group of people. It aims to empower individuals and communities by giving them a voice in representing their realities and shaping decision-making processes.

Social cartography is a participatory method of collective research. It departs from an integrative perspective from which it is understood that reality is constructed culturally by people, from their cultural, interpersonal, and political experiences. All these experiences influence, in turn, on mental, graphic, subjective, and material representations of the socio-cultural context (Photo 3.3).

According to Habegger and Mancila (Habegger et al. 2006), social cartography allows us to know and to build a comprehensive knowledge of the territory to collectively choose a better way of living it, from a better understanding of the territorial reality, of how we live our own territory and of how we build the future territory we want.

From this perspective, we understand social cartography as a process of participatory urban planning in which the actors themselves re-think the territory from the map. They put in the center the knowledge and memories of the people who live in it, and they seek

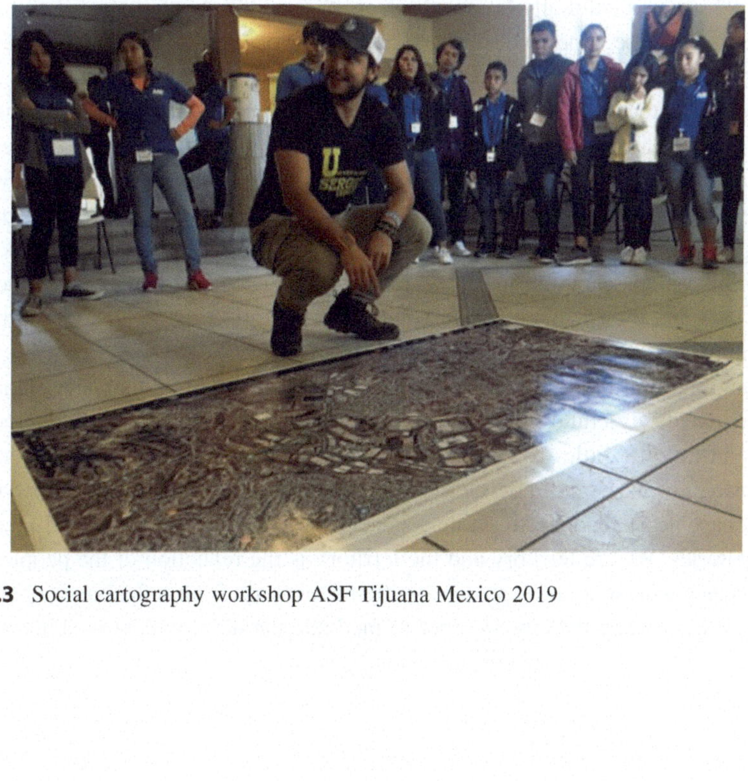

Photo 3.3 Social cartography workshop ASF Tijuana Mexico 2019

collective solutions from citizen participation and from the dialogue between different actors.

According to Oxford University (2021) Social cartography is a method for qualitative research in education. It has been used mostly in comparative and international education, but it has wider applications. It originated from the body of work of Rolland Paulston (1993) who outlined its main conceptual premises and methodological propositions. Unlike other cartographic practices that are mostly concerned with mapping of physical space, social cartography was developed with the purpose of providing a research tool that is capable of mapping relations between and within various epistemic communities and discursive and interpretative frameworks. The practice of social cartography seeks to challenge the positivist and objectivist imperative for singular, "authentic" knowledge and to disrupt the universalizing and totalizing claims of dominant perspectives and frameworks. It does so by mapping the complex and overlapping relations between different discursive and epistemic communities and by situating them in a broader discursive field. Post-representational approaches to social cartography emphasize the agentic properties of maps. These approaches work both on and with the mapmaker and the map reader in ways that seek to trouble and interrupt usual investments in meaning-making and epistemic and discursive privilege

About this workshop

At its simplest, Social Cartography Social cartography is a participatory method of collective research. It departs from an integrative perspective from which it is understood that reality is constructed culturally by people, from their cultural, interpersonal and political experiences. All these experiences influence, in turn, on mental, graphic, subjective and material representations of the socio-cultural context. According to Habegger and Mancila (2006), social cartography allows us to know and to build a comprehensive knowledge of the territory so as to collectively choose a better way of living it, from a better understanding of the territorial reality, of how we live our own territory and of how we build the future territory we want. From this perspective, we understand social cartography as a process of participatory urban planning in which the actors themselves re-think the territory from the map. They put in the center the knowledge and memories of the people who live in it and they seek collective solutions from citizen participation and from the dialogue between different actors.

In urban times our bodies have been disconnected from their natural environment. Our homes have made structural inequalities invisible. Our individuality has broken community ties and little by little we have lost our ability to act for the common good. Given this scenario, it is necessary to develop innovative methodologies from collective research processes capable of reconnecting future generations with the problems and inequalities of our territories. We are territory and the territory is the reflection of the people we are. The experience aims to connect students with the territory, from reflection-action on their own bodies, homes and the city, in order to facilitate the development of skills, aptitudes and capacities for the exercise of the right to the city towards human development and sustainable.

To Run this Workshop:

Number of people: It can be done individually, in groups or just by the decision makers and then presented to other actors in a common meeting.

To share we recommend a 2 h meeting.

Time: 1 day a week or a month depending on the date of the new meeting to prepare

Space: We need a safe dialogue space were each actor or group of actors can safely share their Rich Picture.

Equipment: We need colorful different Sharpies and a big printed image of the physical space that is shared by the different members of the community.

Set up: We will ask the different actors to draw a physical map of their system where does they move to fulfill their role.

We will set a bigger version of that map in the middle of the room and a projection of the map in the wall. We will handle a Sharpie of a different color to each one of them and then ask them to draw the paths where they move in the bigger map. This way they will realize.

3.5 Social Transformation Workshop STW#5 "Viable Systems Model"

STW#5 Viable Systems Model

Organizational Cybernetics Viable Systems Methodology (VSM) St Gallen Management Model (SGM):

The Viable System Methodology (VSM) is an approach developed by Stafford Beer in the field of systems thinking and management. It focuses on understanding and designing organizations as viable systems that can adapt and thrive in complex and changing environments.

After 5 years of applying cybernetics insights to projects, we can establish that there are two different ways in which cybernetics could be applied to project management and the main difference is the Analogy that is privilege the organism analogy or the machine analogy. That way Jackson (2019) differentiate this two approaches "Management Cybernetics or "Organizational Cybernetics" (Fig. 3.4).

In one hand Management cybernetics, the first version is directly related with the mechanical model. Were the goals of an organization are considered unambiguous and are seen as Input-transformation–Output systems like a Flood & Jackson Black Box design to

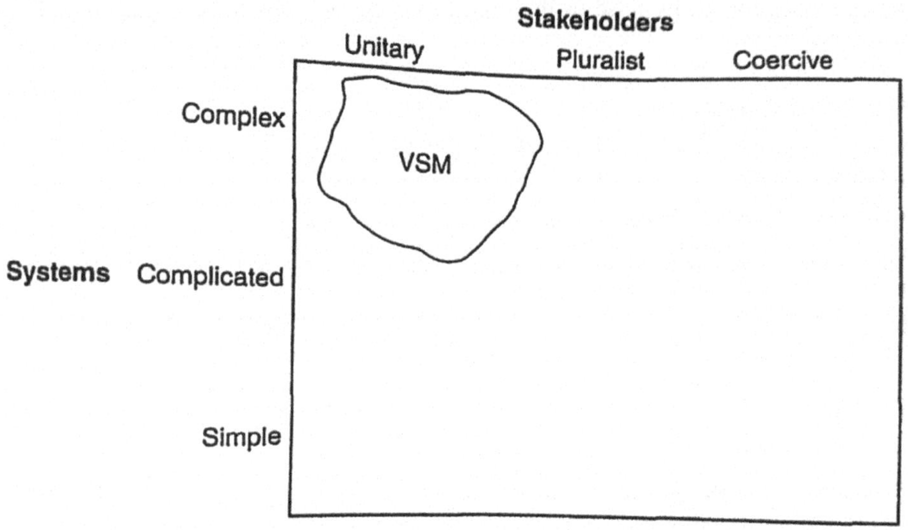

Fig. 3.4 Position of the viable system model on the systems of systems methodology

achieve the main goals efficiently. But this is not the case in most organizations. During the past years the implementation of negative feedback mechanisms to ensure the continuity of projects on time and assure their organizations course were observed. There were different types of feedback not always standardized bud seeking to ensure monitoring and control actions of the different outputs. Trying to assure a Continuous Improvement Cycle as described by the PMI (PMI 2019). On the other hand, what Michael Jackson called Organizational Cybernetics they can transform the way management works. The regulation of an organization in pursuit of some known goals becomes a problem. An organization in a difficult circumstance or a difficult environment will always privilege its own survival instead of the achievement of particular goals. Cybernetics helps us reflect on how organizations and organizations projects could be more adapted to the increasable rapid change and unpredictability of environments these days.

Stafford Beer create a model the viable systems model VSM using almost all cybernetics ideas available at the time. VSM applies easily, coherently, directly and with great acuity to the management of all type of organization. Different authors just like Beer were combined about the need of a generalist integrative and pragmatic approach to management research and education and that this new approach could only by provided by cybernetics and systems theory. One of this Authors was Professor Hans Ulrich from the St Gallen University who created his own version of VSM called the St Gallen Management Model SGM. A first version of the SGM was developed in 1972. It not only employed a stakeholder perspective but also for the first time saw managers and decision makers in social complex systems and organizations as having responsibilities towards the environment!

Other authors inspired by SGM like Scwaninger, Bleicher or Ruegg Sturm attempt to improve this model increasing variety and diversity. A newest version of SGM was released in 2005 (Ruegg-Sturn 2005) VSM and SGM have much in common. For the purpose of this Thesis, we decided to focus on VSM to not duplicate any arguments that may apply to both models. Huhn (2012) and Jackson (2019) states that VSM arrives with a radical break with the traditional managerial model, and the management thinking is more corroborated and has retained its applied focus and relevance to practitioners, while the SGM has tended to become more descriptive in character.

Beer search for naturally occurring systems ruling out mechanical analogies create a cybernetic process, that allows managers to have or at least to believe they have a cybernetic factory a cybernetic organization capable of regeneration, growth and adaptation. As he states

> "*The human body controlled by the nervous system is the richest and most flexible viable system of all.*"
>
> (Beer 1989)

Beer delves into neurophysiology and discovers that there are five essential subsystems, which he labels Systems 1 2 3 4 5 which can be identified in the brain and body in line with mayor functional requirements. This is basis of the VSM, which he then seeks to show is equally relevant to social organizations.

Beer s VSM specifies the criteria that any enterprise must meet if it is to be viable. Meaning capable of surviving and maintaining its identity in an often unpredictable and turbulent environment. These criteria are based on the laws of cybernetics, which beer defines as "*the science of effective organization*" (Jackson 2019).

> "*What truly is given in nature is a collection of principles that systems cannot violate if they are to maintain their identity. These principles are the content of the science of cybernetics, which may conveniently be defined as the science of effective organization.*"
>
> (Beer 1979)

> "*Our institutions are failing because they are disobeying laws of effective organization which their administrators do not know about, to which indeed their cultural mind is closed, because they contend that there exist and can exist no science competent to discover those laws.*"
>
> (Beer 1974)

> "*So far, we have not found any sort of organization to which it does not apply. Just as all types of flying things airliners to Frisbees are subject to the same laws of thermodynamics, so, what it means to be a viable organization seems to be universal.*"
>
> (Hoverstadt 2008)

Managers are confronting with a difficult conundrum; they will be keen to have tight control over their operation s so that the organization is successful in realizing whatever goals it is pursuing. At the same time if they decide to restrict the autonomy of the operations to much then the operations will lack the variety to respond to changes in the environment. The dilemma then presents itself in management theory and practice as the perennial problem of centralization versus decentralization.

> *"Organizations that are too centralized are too rigid, do find it difficult to adapt to changes in their environment and do die as result. Organizations that have no centralizing decision making structures are incapable of acting as coherent wholes and so fall apart."*
>
> (Hoverstadt 2010)

The VSM claims to provide solutions by balancing the variety equations in a way that grants the maximum autonomy possible to the operations that is consistent with overall systemic cohesion.

Stafford Beer, took the first step in Organizational Cybernetics (Beer 1959), that is, in the application of the principles of science Cybernetics to the "**study of organizations**", claiming the system concept as an alternative to the approach reductionist dominant in Western culture and making a defense of holism in the scientific method, just as Aristotle an Capra Before Him he states:

"The whole is more than the sum of its parts",

And so, the whole can only be fully understood if you understand its parts and the nature of their sum. With this systemic approach applied to the study of complex systems, and an organizations, Beer argued for the first time, the possibility of scientifically designing an organization, in such a way that it constitutes a system endowed with the capacities independent existence, regulation, learning, adaptation and evolution necessary to cope with the complexity (Beer 1979) of the environment in which it operates, that is, in a which constitutes a viable system (Perez, Rios 2000–2008). To achieve this, Organizational Cybernetics is based in a series of components, among which, in this work, we will delve into two of them:

1. The Concept of Viability and
2. The Viable Systems Model.

Concept of Viability

The Feasibility Concept is used by Beer to indicate the capacity of an organism an organization a social system, so that, fulfilling the assigned purpose at the same, be able to maintain an independent existence, that is, to survive in the face of changes that may occur in your environment over time even though they were not foreseen when it was designed as it typically happens with engineering artifacts. In In this sense, the purpose of Organizational Cybernetics is ensure the viability of organizations (Beer 1985).

Model of Viable Systems

The Viable Systems Model is one of the main components of Organizational Cybernetics and one of the best known and most used contributions of Beer in the scope of Organization Theory (Beer 1972, 1979, 1981, 1985). The Advantages of its use lie in its systemic nature, comprehensive and multilevel, as well as its capacity for the management of the dynamic complexity that the management of an organization and its interaction with the environment.

It establishes the necessary and sufficient conditions that every organization must comply to be a system viable. Specifically, the presence and proper functioning of a series of functions or subsystems identified as essential in any organization (System One, System Two, System Three, System Four and System Five) and one series of communication relationships between these functions or subsystems or, between them and the environment in which operate [Fig. 2.1].

System One

Sistema One is responsible for producing and delivering to the environment the goods or services that the organization Produces. It is therefore constituted by the processes productive or the operational elements that make it possible that the organization generates said goods or services. Every one of these operational elements will be a system viable complete, with a separate existence within the organization, that is, with its environment, nature, purpose, and particular responsibilities. In Fig. 2.1 we can see an example in which the System.

One is made up of two operational elements 4. Each one of these operational elements is composed of a fundamental unit of operation represented by the green circle, a fundamental management unit represented by the green square and an environment by which it is affected Represented by the gray Shape or amoeba. The operational elements are connected to each other by means of fundamental units of operation. The rest of the Systems, from Two to Five, have as mission to serve System One, that is, to contribute to it fulfill its purpose that is the delivery of goods or services to the environment.

System Two

System Two is intended to achieve smooth operation harmonic for all operational elements that make up System One. To do this, it will have to cushion oscillations that occur as a consequence of their autonomous functioning and of the interactions that they occur between them. Each operational element has a System Two local represented by the green triangles with top vertex, they give information regarding the operation of the rest of operational elements in order to coordinate their performances and interactions. All Systems Two premises are connected to the corporate System Two represented by the yellow triangle with upper vertex that feeds the System Three with information on the progress of the operational and sense elements. On the contrary, it transmits to the elements operational the necessary information to coordinate their activities. To achieve such coordination, the System Two uses mechanisms and tools that make it happen practically automatically and if possible, without the need for intervention of any kind of authority, like System Three.

System Three

Sistema Three is in charge of optimizing the operation of the whole System One composed of the different operational elements. Is he System that has a vision of the whole that none of the elements operational operations in isolation owns and the only trained to identify integration and potential synergies that may appear between elements operational that make up the System One It can be said that the main System Three's role is to take care of that everything works correctly "*here and now*", of the present of the organization, monitoring its operation in the short term. System Three, therefore, will try to maintain internal stability of the organization, taking care of the internal real-time environment. For this, System Three represented by the square lower yellow is related to operational elements of the System One to:

(1) Transmit information, instructions, goals, objectives and guidelines from System Four and System Five up to System One. Definitive, to transmit information Fig. 2.1: Viable Systems Model (Beer 1985) on the way of acting in general of the organization as a whole. In the opposite direction, also channels relevant information regarding operation from System One to System Four and System Five (Figs. 3.5 and 3.6);
(2) Manage the use and distribution of resources available with the operational elements of the System One and
(3) Follow the operation of these through the accountability they have to perform (reporting on the progress of operations, fulfillment of objectives and additional resource needs). In short, the System Three controls the performance of the operational elements. Other functionality of System Three appears in the case that the conflicts arise between the elements that make up System One, they have not been able to be solved through the interaction between the operational elements or through the mechanisms of coordination or System Two. In this case, in addition to extraordinary circumstances that require it, the System Three will intervene directly using their authority. Finally, supporting System Three appears the System Three * represented by the yellow triangle with lower vertex, whose main mission is to obtain information about the System One that is not obtained through the direct relationship between System Three and System One or through System Two. It is the person in charge, therefore, of providing information directly from operational elements, avoiding relying on the information they send to System Three. It allows audit the operation of System One by representing control and review mechanisms, both internal and external.

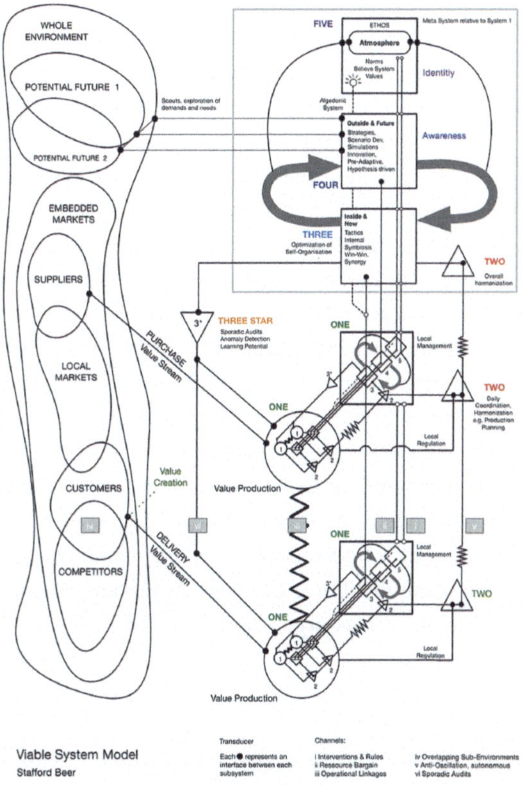

Fig. 3.5 Viable systems model staford–beer, S. diagnosing the system for organizations

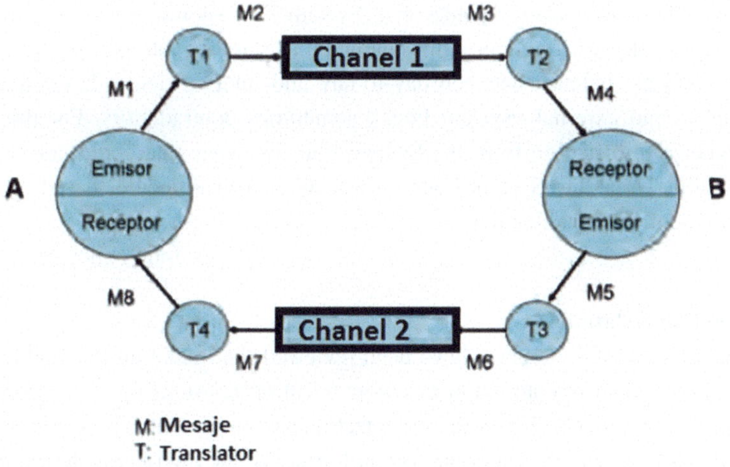

Fig. 3.6 Elements from a communication relation

System Four

System Three is not capable of acting in anticipation of the future, is unable to recognize the potential risks that may appear. We need a structural function that solves this inconvenience. This function is System Four represented by the middle yellow square, in charge of monitoring the evolution of the organization's environment. Its main mission is to deal with the "*exterior and future*" of the organization, with the purpose of keeping it constantly prepared for the change. In addition to monitoring the environment at the moment current, System Four should try to obtain information on anticipated moments or possible situations for the future. All the news identified by System Four as necessary for the organization, they must be understood and accepted by System Three and consequently implanted in System One, thus ensuring adaptability of the organization. To complete this process of adaptability, System Four will receive from System Three information about what is happening in the present within the organization itself as well as on the capacity of System One to incorporate the proposed changes by System Four. In the same way, the System Four will send to System Five the information relevant to this one. Among other elements, System Four will ideally be formed by the "*operating room*", where they are explored continuously different future scenarios to help make decisions that increase the probability to achieve the desired future. Changes in the environment detected are analyzed with reference to the objectives organization, leading to possible recommendations of performance.

System Five

Finally, System Five represented by the upper yellow square deals with the ideological and normative aspects of organization, defining its mission, global objectives, style and identity and establishing its character, that is, allowing express what you want the organization to be, how it is you want it to behave internally and how you want it to the organization is perceived from outside of it. System Five should ensure that the organization is adapt to the environment while maintaining stability internal. It must therefore achieve an adequate balance between day-to-day and internal demands (System Three) with the future and external (System Four), sometimes contradictory. For this, System Five will receive information from the System four on issues that could not be resolved between System Three and System Four, so that System Five decides based on ideology and regulations of the organization.

John Wiley & Sons. 1985.

Communication Relations

Once the main structural elements have been reviewed that make up the Viable Systems Model, is important to pay attention to communication relationships that connect these elements with each other and with the organization's environment. These communication relationships seek to achieve a continuous equilibrium in the interaction between the connecting elements. The balance is achieved when the information in quantity, in content, in format and in time is adequate so that the relationship between both elements is a dynamic but harmonious relationship, in the sense that it works the way you want by both parties (Perez Rios 2000–2008) Among them, there is one of special importance which we call the alarm channel or algedonic channel connecting operational elements from System One directly to System Five. East the alarm channel has the function of alerting System Five of any circumstance that has occurred in the System One, which could not be resolved by him or by him System Three, and that may pose a serious risk to the viability or operation of the organization. Eight intervening in these communication relationships elements. The emitter (A), the "transducer" (T1) that encodes the information that comes out of the issuer to be able to be transported, the communication channel (Channel 1) that must be able to drive the amount of information per unit time required, again "transducer" (T2) that decodes the information coming from the communication channel and converts it to the proper format so that the interpret the receiver (B). In the reverse way, the receiver and the issuer begin to exchange their roles, finding again the same elements but in reverse [6].

Recursive Character

Another essential aspect of the Viable Systems Model is its recursive character. The Principle of Recursion another component of Organizational Cybernetics, states that every viable system contains viable systems, and, in turn, it is part of systems that are also viable. In Fig. 2.1 you can see how inside the green circles and squares that represent the operational elements of the System One is contained an exact replica of the structure of the Model. The direct consequence of this recursion is that any viable system, regardless of the level of recursion it occupies, it must contain the five subsystems o functions defined by the Model, that is, the feasibility of the system requires that all five functions exist, recursively, at all levels of the organization.

The complexity of the environment in which they develop their organizations' activity, makes it increasingly necessary use approaches that have sufficient capacity to face it. The approach or systems thinking offers a wide variety of methodologies, models, particularly suitable methods, tools, and techniques for this end. Among them we highlight Cybernetics Organizational and particularly two of its components: the Feasibility Concept and the Viable Systems Model. The Viable Systems Model offers us the possibility to design (if it is new) or diagnose (in case it is new) the case that an organization already exists) scientifically, so that it constitutes a system endowed with the capabilities independent existence, regulation, learning, adaptation and evolution necessary to cope with the complexity of the environment in which it operates, thus constituting a viable system. The design or structural diagnosis of the organization will be determined by the way in which it divides the complexity of the environment in which it operates and configures accordingly the different levels of recursion needed to cope with it, as well as the number of operational elements that we will find in each level. Any lack either due to the lack of existence of some level of recursion, some subsystem, due to the malfunction of subsystems existing or due to a non-existent or deficient design of communication relationships will lead to pathology corresponding organizational structure, which will make the organization not work well or even disappear. Generally, these pathologies can be classified into three basic groups: (a) structural pathologies, (b) functional pathologies and (c) information-related or "informational" pathologies. The detail of these pathologies can be reviewed in (Perez Rios 2000–2008). The validity of this Model is justified by different quantitative studies and by a large number of applications of a qualitative nature. Industry steel, textile industry, insurance companies, banks, railways, shipping companies, universities, health, and government at the local, regional, and state have applied with successful results the guidelines set by the Viable Systems Model. Among all of them, the application of the Model in the system Chilean politician in the 70 s is the most remarkable. May see a complete description of this application in Beer (1981). Despite this, one of the main criticisms made on the Model is based on its scarce empirical validity. This, along with other criticisms such as the minimization of the importance of individuals in organizations, their character abstract, its scarce formalization, and the lack of clear procedures for its application encourage us to follow deepening in the Viable Systems Model and by extension in Organizational Cybernetics with the objective to understand the treatment that this science makes on these criticisms.

The most important thing for any viable system is not that it possesses a scientific model of its environment but that it is set up appropriately to respond to things it has not encountered before. Successful adaptation becomes then the most significant Goal. Beer´s VSM is important because it shows how *"a dance of becoming"* can be established between an organization and its environment during which they explore each other´s possibilities, establishing mutually satisfactory relationships along the way. They evolve together continuously, and it is impossible to predict what will happen. We simply have to watch where the performance staged by the VSM take us. Knowledge is not fixed but emerges as system and environment find out what each other are capable of.

As Heidegger States, *"The world is revealed rather than enframed"*.

The VSM provided a way of organizing the different levels of the industrial economy in a manner that would allow them to learn their way to increased performance, under workers control, while adapting to unpredictable disturbances… (Pickering 2009)

But there was not detailed scientific plan setting out how the industrial economy should achieve its goals. More importantly, because he was there Raul Espejo also agrees with Pickering afirmations, seeing VSM Allende's third way intervention in Chile as Performative.

> Beers work bears the hallmarks of British cybernetics as described by Pickering … Beer was more interested in studying how systems behaved in the real world than in creating exact representations of how they functioned… He stressed that cybernetics and operations research should drive action, not to create mathematical models of increasing complexity and exactitude ….Instead of using science to dominate the outside world, scientist should focus on

> identifying the equilibrium conditions among subsystems and developing regulators to help the overall system reach its natural state of stability
>
> (Espejo and Reyes 2014).

> VSM acts as a hermeneutic enabler by setting out the structural context in which good conversations and problem solving can take place
>
> (Espejo and Reyes 2011).

> The VSM is above all about enabling connectivity and structuring the system to facilitate the healthy development of relationships and ultimately effective performance
>
> (Espejo and Reyes 2011).

Beer was determined to break with traditional management thinking. He looked at companies and their organization charts and despaired. To him they were totally unsatisfactory as a model for complex enterprises. They suggested that the person at the top needed a brain weighing half a ton-since all information flowed up to him and all decisions appeared to be his responsibility. As Beer commented, people´s heads do not get bigger towards the top of an organization, except perhaps in a metaphorical sense. Starting with neurophysiological analogies and drawing upon cybernetics, the "science of effective organization "he set out to construct a more useful model. The result was his influential VSM model a model of the key features that any viable systems must exhibit. This provides a radical alternative to "mechanistic" thinking and suggests organizational forms that are more appropriate in today´s turbulent environments. Furthermore, the VSM conveys its insights in a relatively useable form. Managers just need to apply the model. They have only themselves to blame if they don´t try to understand it and the lessons it imparts. They will face difficulties. All systems models and methodologies struggle with the complexity of today´s problems situations. But they can rest assured that they are better and more relevant to practice than those produced by any other discipline.

About this workshop

At its simplest, The Viable Systems Workshop allow us to evaluate the SCS structure to see if it is able to survive, meaning that it is homeostatic balanced both internally and externally and has opportunities to grow and to learn to evolve and to adapt: To "become more and more potent in its environment" (Beer 1981)

Beer specifies the criteria that any enterprise must meet if it is to be viable: it defines the laws of cybernetics known as the science of effective organizations.

Organizations are too centralized to rigid, finding difficult to adapt to change in the environment and do die as a result. Organizations that have no centralizing decision making structures are incapable of acting as coherent wholes and do fall apart (Hoverstadt 2010 = Viable systems are made of 5 interconnected sub-systems each one with a specific function).

Institutions that fail are probably disobeying the laws of effective organization which their administrator do not know about.

To Run this Workshop:

Number of people: It can be done individually, in groups or just by the decision makers and then presented to other actors in a common meeting.

Time: To share the results and debrief we recommend 2 h meeting it could be 1 day a week or a month after the other workshop depending on the date of the new meeting to prepare.

Space: We need a safe dialogue space were each actor or group of actors can safely share the subsystems they have identified..

Equipment: We can use a printed version of the common system or a digital one it is important that we can easily add or remove elements from the model.

Set up: We will previously ask the actors to try to identify possible subsystems within the SCS and see if they can identify which subsystems are responsible of the 5 main functions (Implementation, Co-ordination, Operational control, Auditing, Development and Policy).

Then in a safe dialogue space we will ask the different participants to see if they were able to identify all the 5 systems. If any of those systems was not Identify we will ask them why they believe they are missing that specific system and what they think they can do about it.

System 1 Implementation responsible for producing and delivering the result. System One is responsible for producing and delivering to the environment the goods or services that the organization produces. It is therefore made up of the production processes (operational elements) that make it possible for the organization to generate said goods or services. Each of these operational elements will be a complete viable system, with a separate existence within the organization, that is, with its particular environment, nature, purpose and responsibilities.

Is there a system 1 in your system if so which one? Otherwise, what is needed to implement it?

Represent subsystem 1:

System 2 Coordination aims to achieve a harmonious operation for the set of operational elements that make up System One. To do this, it will have to dampen the oscillations that occur as a result of its autonomous operation and the interactions that occur between them. Each operational element has a local System Two (green triangles with an upper vertex) that provides information regarding the operation of the rest of the operational elements in order to coordinate their operations and interactions. All the local Two Systems are connected to the corporate System Two (yellow triangle with upper vertex) that feeds System Three (Section 3.3) with information on the progress of the operational elements and, conversely, transmits to the operational elements the information necessary to coordinate their activities. To achieve such coordination, the System Two uses mechanisms and tools that make it occur practically automatically and, if possible, without the intervention of any type of authority (System Three).

Represent subsystem 2:

System 3 Control System, System Three is in charge of optimizing the operation of the whole of System One made up of the different operational elements. It is the System that has a vision of the whole, that none of the operational elements in isolation has and the only one capable of identifying the integration and potential synergies that may appear between the operational elements that make up System One. It can be said that the main function of the System Three is to ensure that everything works correctly "here and now", in the present of the organization, monitoring its operation in the short term. System Three, therefore, will try to maintain the internal stability of the organization, dealing with the internal environment in real time. For this, System Three (lower yellow square) is related to the operational elements of System One to: (1) transmit information, instructions, goals, objectives, and guidelines from System Four and System Five up to System One. In short, to transmit information about the organization as a whole. On the contrary, it also channels the pertinent information regarding the operation of System One up to System Four and System Five; (2) manage the use and distribution of available resources with the operational elements of the System One and (3) follow their operation through the accountability they have to carry out (reporting on the progress of operations, meeting objectives and needs for additional resources). In short, System Three controls the performance of the operational elements. Other functionality of System Three appears in the event that the conflicts arise between the elements.

Operational elements that make up System One, could not be resolved through the interaction between the operational elements themselves or through the coordination mechanisms (System Two). In this case, in addition to extraordinary circumstances that require it, System Three will intervene directly using its authority.

Represent subsystem 3

System 3*. Supporting System Three appears System Three * (yellow triangle with lower vertex), whose main mission is to obtain information about System One that is not obtained through the direct relationship between System Three and System One or through the System Two. It is therefore in charge of providing information directly from the operational elements, avoiding relying on the information that they send to System Three. It allows auditing of the operation of System One representing internal and external control and review mechanisms.

Represent subsystem 3*

System 4 Intelligence System: System Three is not capable of foreseeing the future, it is not capable of recognizing the potential risks that may appear. We need a structural function that solves this problem. This function is System Four (intermediate yellow square), in charge of monitoring the evolution of the organization's environment. Its main mission is to deal with the "exterior and future" of the organization, with.

The purpose of keeping it constantly prepared for change. In addition to monitoring the environment at the present time, System Four should try to obtain information on anticipated moments or possible situations for the future. All the innovations identified by System Four as necessary for the organization, must be understood and accepted by System Three and consequently.

Implemented in System One, thus ensuring the adaptability of the organization. To complete this adaptability process, System Four will receive from System Three information about what is currently happening within the organization itself as well as about System One's ability to incorporate the changes proposed by System Four. In the same way, System Four will send System Five (Section 3.5) the relevant information for it. Among other elements, System Four will ideally consist of the "operating room", where different future scenarios are continuously explored to help decision-making that increases the probability of achieving the desired future. The detected environmental changes are analyzed with reference to the main objectives of the organization, leading to possible recommendations for action.

Represent subsystem 4

System 5 Political Function Finally, System Five (upper yellow square) deals with the ideological and normative aspects of the organization, defining its mission, global objectives, style and identity and establishing its character, that is, allowing it to capture what it wants to be the organization, how you want it to behave internally and how you want the organization to be perceived from outside it. System Five must ensure that the organization adapts to the environment while maintaining internal stability. It must therefore achieve an adequate balance between day-to-day and internal demands (System Three) with future and external demands (System Four), sometimes contradictory. For this, System Five will receive information from System Four on the issues that could not be resolved between System Three and System Four, so that System Five decides based on the ideology and regulations of the organization.

Represent Sub-System 5

	CATWOE
Clients	
Actors	
Transformation	
Worldview- Welstangchauung	
Decision makers -owners	
Environmental limits	

Each actor chooses one of the identified subsystems to make a Catwoe of the system the Subsystem.

Each actor makes a Root definition of the identified subsystem.

The system does X through Y to achieve Z.

Communication Relations

Having reviewed the main structural elements that make up the Viable Systems Model, in this section we will pay attention to the communication relationships that connect these elements with each other and with the organization's environment. These communication relationships seek to achieve a continuous balance in the interaction between the elements that connect. The balance is achieved when the information (in quantity, content, format, and time) is adequate so that the relationship between the two elements is a dynamic but harmonic relationship, in the sense that it works in the way desired by the two parts [6]. Among them, there is one of special importance, which we call the alarm channel or algedonic channel that connects the operational elements of System One directly with System Five. This

The function of the alarm channel is to alert System Five of any circumstance that may have occurred in System One, which has not been resolved by System One or by System Three, and which may pose a serious risk to the viability or operation of the organization. Eight elements intervene in these communication relationships [Fig. 2.2]. The emitter (A), the "transducer" (T1) that encodes the information that leaves the emitter in order to be transported, the communication channel (Channel 1) that must be able to conduct the amount of information per unit of time that is again requires a "transducer" (T2) that decodes the information from the communication channel and converts it to the appropriate format for the receiver (B) to interpret. In the reverse way, the receiver and the emitter begin to exchange their roles, finding the same elements again but in the reverse way [6].

Another essential aspect of the Viable Systems Model is its recursive nature. The Principle of Recursion (another component of Organizational Cybernetics) states that every viable system contains viable systems and, in turn, is part of systems that are also viable. In Fig. 2.1 it can be seen how within the green circles and squares that represent the operational elements of System One, an exact replica of the structure of the Model is contained. The direct consequence of this recursion is that any viable system, regardless of the recursion level it occupies, must contain the five subsystems or functions defined by the Model, that is, the viability of the system requires that the five functions exist, recursively, at all levels of the organization.

3.6 Social Transformation Workshop STW#6 "Organizational Pathologies"

STW#6 Organizational Pathologies

Organizational Pathologies

Organizational pathologies refer to dysfunctional or problematic patterns and behaviors that can emerge within an organization. These pathologies can hinder the organization's effectiveness, productivity, and overall health. Here are some common organizational pathologies:

Having considered basic cybernetic principles and the main elements of the VSM, the way in which they can be applied in diagnosing or designing organizations, I will now turn to some of the pathologies which most commonly occur in organizations or companies (see Perez Rios 2008).

As we know from medical practice, detecting a pathology is essential when it comes to prescribing the treatment required for the diagnosed deficiency. In the case of organizations, a knowledge of all the most frequent pathologies, the individual characteristics of each one and the indications for dealing with them is in itself very useful for the managers responsible, as it enables them, on the one hand, to identify the problem rapidly and, on the other, to decide on how it might be dealt with. One class of diagnostic examples found in other areas of Systems Thinking employs common structures called "*archetypes*" (Senge 1990), which facilitate conversation among the managers involved to speed up the process of identifying the particular problem. In the case of OC, there are also studies with a similar aim (Beer 1989; Schwaninger 2001; Espejo 2008).

In order to list the diverse potential pathologies, a conceptual outline was use concerning the diagnosis and design of organizations. First of all, we will see pathologies related to what we can consider a vertical dimension, that is to say, those concerning the organization's structural design in relation to the general environment and the various environments which it contains. I will call those encompassed here Structural Pathologies.

Next, we will look at those related to the adequacy of organizations at all recursion levels, to the prescription made by the VSM about functional subsystems and their relationships. These pathologies will be called Functional Pathologies. Finally, the third group subsumes information systems and communication channels pathologies.

Structural Pathologies

The pathologies included in this group are related to an inadequate treatment of the total complexity faced by an organization. The organization and its relevant environmental size may indicate a division of the environment into sub-environments, and the same with the organization. This vertical unfolding of complexity enables it to comply with Ashby's law, each sub-organization having to deal with less complexity or variety. Dividing up the environment into smaller environments and assigning these to the corresponding organizations means that the latter are faced with less complexity. The pathologies normally showing up in this context are concerned with either the absence of this process of division when in fact it is necessary, the absence of organizations corresponding to some particular level environment, or confused organizational membership's multiple-dependence relationships. The pathologies identified in this group are:

(1) Non-existence of vertical unfolding,
(2) Lack of recursion levels (first level),
(3) Lack of recursion levels (middle levels) and
(4) Entangled vertical unfolding with various interrelated level memberships.

Functional Pathologies (VSM)

Having analyzed certain pathologies associated with vertical partitioning of complexity and the manner in which this is distributed throughout different organizations at different recursion levels, we will now examine some of the common pathologies relating to each of the organizations that compose the total organization, which I call Functional Pathologies. The aim is to see whether the essential functions systems necessary for the organization's viability exist and work adequately.

Pathologies Associated with Information Systems and Communication Channels

Communication channels within the VSM they work as elements connecting all the functions/subsystems of the organization, and the persons who compose it, as well as the organization with the different parts of the environment. Furthermore, mention was made of the requirements needing to be met by a channel, if it is to suitably fulfill its function as a transmitter of information. Therefore, what I intend to do in this section is to point out certain possible pathologies associated with the existence and constitution of communication channels and, in wider terms, to the information systems.

Design and Diagnosis of Organizations by Means of the VSM

Applying Beer's VSM to both the design and diagnosis of organizations makes it possible, on the one hand, to guarantee that they will be equipped with the necessary and sufficient elements required to ensure their viability and, on the other, to identify which elements are not embodied as they should be, whether they function poorly, or whether, despite having the necessary structural components for smooth operation at least initially, there is a dysfunction on account of deficiencies associated with the information systems and communication channels. The most common pathologies in organizations, classified in three main groups and each corresponding to one of the essential dimensions of OC application. The first group includes Structural Pathologies; in the second Functional Pathologies; and in the third group we find Pathologies associated with information systems and communication channels. The availability of this taxonomy of pathologies may, besides providing certain keys for identifying and subsequently correcting commonly occurring problems in organizations, also be particularly useful for helping managers to apply the VSM in organizations. One of the reasons highlighted by various researchers as possibly accounting for the relatively slow process of adoption and application of OC in companies and organizations in general, is, as I have already indicated, the difficulty experienced in understanding and applying it. The availability of new technological tools such as VSM model software (Pérez Ríos 2003, 2006, 2008) combined with the introduction of application guidelines may, I believe, be useful in terms of both understanding the OC/VSM methodology and applying it.

About this workshop

At its simplest, The Organizational Pathologies Workshop seeks use certain basic cybernetic principles and the main elements of the VSM, to diagnose possible organizational pathologies which most commonly occur in organizations or companies (Perez Rιos 2008).

As we know from medical practice, detecting a pathology is essential when it comes to prescribing the treatment required for the diagnosed deficiency. In the case of organizations, a knowledge of all the most frequent pathologies, the individual characteristics of each one and the indications for dealing with them is in itself very useful for the managers responsible, as it enables them, on the one hand, to identify the problem rapidly and, on the other, to decide on how it might be dealt with. One class of diagnostic examples found in other areas of Systems Thinking employs common structures called "archetypes" (Senge 1990), which facilitate conversation among the managers involved to speed up the process of identifying the particular problem.

There are three types of organizational pathologies first of all, we will see pathologies related to what we can consider a vertical dimension, that is to say, those concerning the organization's structural design in relation to the general environment and the various environments which it contains.

I will call those encompassed here Structural Pathologies. Next, we will look at those related to the adequacy of organizations (at all recursion levels) to the prescription made by the VSM about functional subsystems and their relationships. I call these Functional Pathologies. Finally, the third group subsumes information systems and communication channels pathologies.

Number of people: It can be done individually, in groups or just by the decision makers and then presented to other actors in a common meeting.

Time: To share the results and debrief we recommend a 2 h meeting it could be 1 day a week or a month after the other workshop depending on the date of the new meeting to prepare.

Space: We need a safe dialogue space were each actor or group of actors can safely share the subsystems they have identified and those that are missing and the reasons why they believe this is happening.

Equipment: We can use a printed versions of the subsystems identified or a digital one it is important that we can easily add or remove elements from the models.

Set up: We previously ask the actors to try to identify possible subsystems within the SCS and see if they can identify which subsystems are responsible of the 5 main functions (Implementation, Co-ordination, and Operational control, Auditing, Development and Policy)

Now we will ask them to try to identify possible pathologies based on the previous workshop.

Organizational pathologies identified:

Pathologie	Símptoms	Proposed treatment actions

3.7 Social Transformation Workshop STW#7 "Systems Archetypes"

STW#7 Systems Arquetypes- Arquetipos Sistémicos

Systems Arquetypes

Systems archetypes are recurring patterns or structures that are commonly observed in complex systems. These archetypes help in understanding and diagnosing the behavior of systems and identifying potential leverage points for intervention. Here are some well-known systems archetypes:

The Systems Archetypes describe common patterns of behavior in organizations. As diagnostic tools they provide insight into the underlying structures from which behavior over time and discreet events emerge. As prospective tools, they alert managers to future unintended consequences. Collectively they challenge managers to consider the merits of fundamental solutions by making time an explicit variable in decision making.

The System Archetypes are highly effective tools for gaining insight into patterns of behavior, themselves reflective of the underlying structure of the system being studied.

The archetypes can be applied in two ways-diagnostically and prospectively.

1. Diagnostically, archetypes help managers recognize patterns of behavior that are already present in their organizations. They serve as the means for gaining insight into the underlying systems structures from which the archetypal behavior emerges. This is the most common use of the archetype.

Archetypes are effective tools for beginning to answer the question,

"Why do we keep seeing the same problems recur over time?"

Archetypes are also useful prospectively for planning. As managers formulate the means by which they expect to accomplish their organizational ends, the archetypes can be applied to test whether policies and structures under consideration may be altering the organizational structure in such manner as to produce the archetypal behavior. If managers find this to be the case, they can take remedial action before the changes are adopted and embedded in the organization's structure.

Where did archetypes come from?

In the 1960s and 1970s, Jay Forrester, Dennis Meadows, Donella Meadows, and other pioneers of systems thinking observed several recurring systemic structures. In the 1980s, Michael Goodman, Charles Kiefer,

Jenny Kemeny, and Peter Senge built on that work, in part with the help of notes developed by John Sterman, by describing, diagramming, and cataloguing these generic systemic structures as systems templates. When Peter Senge authored The Fifth Discipline: The Art and Practice of the Learning Organization, he referred to those structures as systems archetypes.

Since then, the notion of systems archetypes has become quite popularized, and systems thinking practitioners have continued to teach, apply, and write about these recurring generic structures as well as investigate and test the potential of identifying new ones.

Archetypes and Modeling

Archetype are useful for gaining insight into the "nature" of the underlying problem and for offering a basic structure or foundation upon which a model can be further developed and constructed. The archetypes are rarely sufficient models in and of themselves. They are generic in nature and generally fail to reveal important variables that are part of the real system structure of a specific organization. Without an explicit awareness of these real variables, it is difficult for managers to pinpoint specific leverage points where changes in structure can achieve sustainable changes in system behavior.

The word comes from the Greek Archetypes, which means "primitive or original model". A stepson of systems thinking systems archetypes were developed at Innovation Associates in the mid-1980s. At that time, the study of system dynamics depended on the diagramming of complex causal diagrams and computer modeling, using mathematical equations to define the relationships between variables. Charles Kiefer, the President of I.A., suggested trying to communicate the concepts more simply. Jennifer Kemeny (with Michael Goodman and Peter Senge, based in part on the notes developed by John Sterman) developed eight diagrams that would help illustrate the most commonly seen behaviors. Some archetypes, including "Limits to Growth" and "Load Displacement" were translations of "generic structures," mechanisms that Jay Forrester and other pioneers of systems thinking described in the 1960s and 1970s.

Most known Archetypes:

1. Compensation between process and delay
2. Limits of growth
3. Shifting the load
4. Special case: shifting the load towards the intervention
5. Erosion of goals
6. Climbing
7. Success for those who success
8. Tragedy of the common ground
9. Quick solutions that fail
10. Growth and underinvestment.

How can you use the archetypes?

When you are trying to examine an issue, a problem, or a situation systemically, the numerous tools of systems thinking can help you in your inquiry. The systems archetypes, as one class of systems thinking tools, can open a whole new dimension of your learning. Because each archetype comes with its own storyline (or causal theory), pattern of behavior over time, and systemic structure depicted in a causal loop diagram (like templates), you can connect to it and apply its learnings from any of those entry points (see "A Review of Causal Loop Diagrams"). As you work on a problem—going over the story, identifying key variables, sketching causal loop diagrams—you may suddenly find that something about the story resonates with one or more of the archetypes, giving you a new key for unlocking part or all of your problem. At other moments, you may finish graphing the behavior over time of key variables, and notice a pattern that seems to match a behavior pattern associated with one of the archetypes. A rich vein of questions, theories, and possibilities opens up for your investigation. Or, you may be looking over a causal loop diagram drawn by a colleague and notice a particular combination of loops that reminds you of the core structure of a certain archetype, and that prompts you to ask a question that had not yet occurred to either of you. At first, you'll probably find your own way of applying your understanding of the system archetypes—through using story themes, structure- behavior pairs, or generic causal loop diagrams. Eventually, you can use all three aspects of the archetypes to broaden your perspective on systemic problems, generate additional or unexpected questions, notice when you are experiencing one or more of these recurring dynamics, and anticipate possible future outcomes of current actions and events.

Connections between the Archetypes

There are many ways in which the archetypes can interact with each other. Michael Goodman and Art Klien have mapped the archetypes, and their interactions, in an article published in The Systems Thinker 20 in the December, 1993/January, 1994 issue. The work was subsequently republished in the Fifth Discipline Fieldbook 21, by Peter Senge et al.

Conclusion About Systems Arquetypes

The Systems Archetypes are patterns of behavior that emerge from the underlying system structure. They can be used diagnostically to reveal insights into the structure that already exists, or prospectively to anticipate potential problems and/or problem symptoms.

Archetypes do not describe any one problem specifically. They describe families of problems generically. Their value comes from the insights they offer into the dynamic interaction of complex systems.

Use them with the other systems thinking tools that are available. As part of a suite of tools, they are extremely valuable in developing broad understandings about organizations and their environments and contribute to more effectively understanding problems.

> We don't need better solutions; we need better thinking about problems.
>
> (Ackoff 1991).

If managers can learn to recognize systems archetypes, they can save themselves a lot of wasted effort and target their interventions at points of maximum leverage (Figs. 3.7, 3.8 and 3.9).

Fixes that Fail: In this archetype, a problem is addressed using a solution that initially appears successful. However, over time, the solution produces unintended consequences that worsen the original problem or create new ones. The attempted fix ultimately fails to address the underlying dynamics and may exacerbate the situation.

Shifting the Burden to the Intervenor: This archetype involves a situation where the responsibility for problem-solving is shifted to an external intervenor or authority, relieving the primary actors from their own responsibility. The intervenor takes on the burden of addressing the problem, which can create dependency and prevent the primary actors from developing their own capacity to solve problems.

At it simplest, the Systems Archetypes workshop seeks to describe common patterns of behavior in organizations.

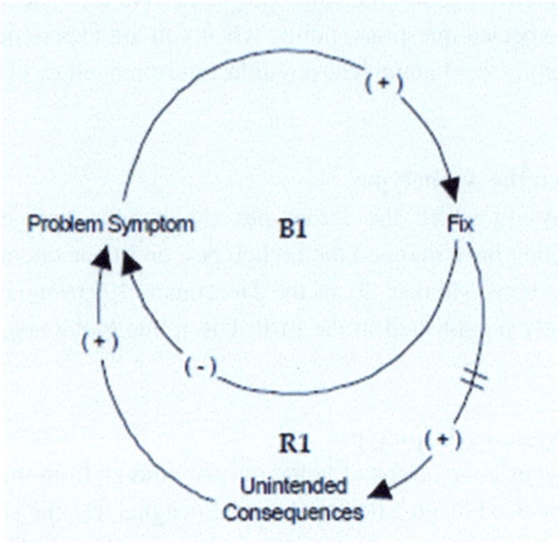

Fig. 3.7 Fixes that fails system archetypes (Braun 2002)

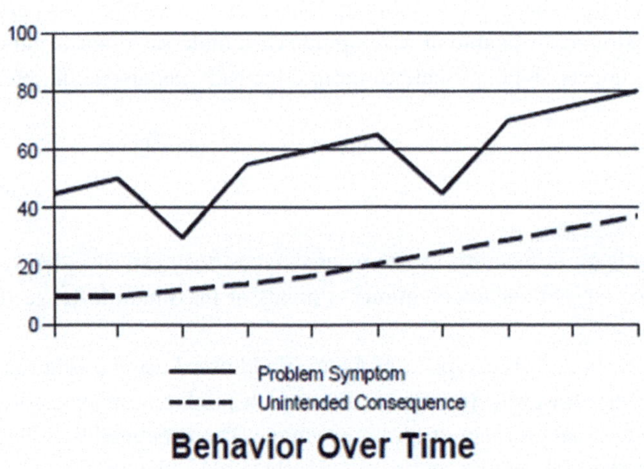

Fig. 3.8 Fixes that fails system archetypes (Braun 2002)

As a diagnostic tool they provide insight into the underlying structures from which behavior over time and discreet events emerge.

As prospective tools, they alert managers to future unintended consequences. Collectively they challenge managers to consider the merits of fundamental solutions by making time an explicit variable in decision making.

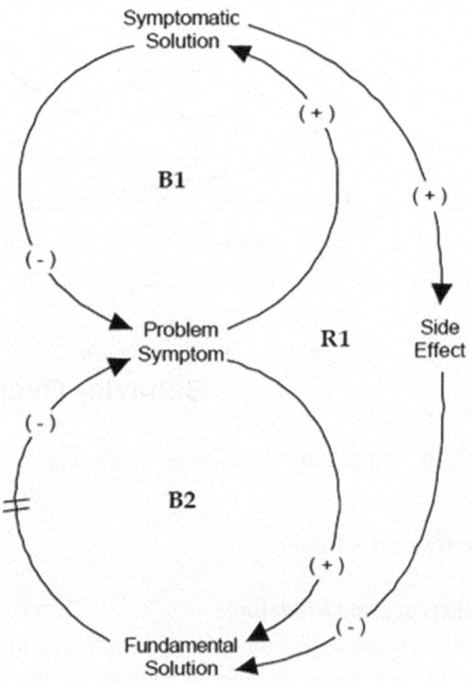

Fig. 3.9 Shifting the burden system archetypes (Braun 2002)

The System Archetypes are highly effective tools for gaining insight into patterns of behavior, themselves reflective of the underlying structure of the system being studied. The archetypes can be applied in two ways - diagnostically and prospectively. Diagnostically, archetypes help managers recognize patterns of behavior that are already present in their organizations. They serve as the means for gaining insight into the underlying systems structures from which the archetypal behavior emerges. This is the most common use of the Archetype.

Archetypes are effective tools for beginning to answer the question, "Why do we keep seeing the same problems recur over time?" Archetypes are also useful prospectively for planning. As managers formulate the means by which they expect to accomplish their organizational ends, the archetypes can be applied to test whether policies and structures under consideration may be altering the organizational structure in such manner as to produce the archetypal behavior. If managers find this to be the case, they can take remedial action before the changes are adopted and embedded in the organization's structure (Fig. 3.10).

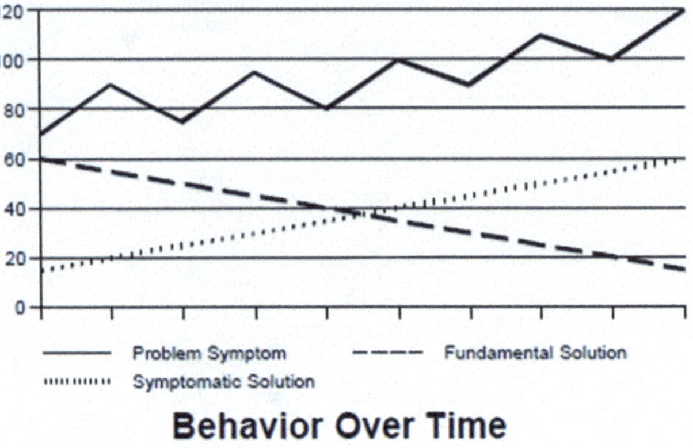

Fig. 3.10 Shifthing the burden system archetypes behavior over time (Braun 2002)

About this workshop:

Archetypes and Modeling

Archetypes are useful for gaining insight into the "nature" of the underlying problem and for offering a basic structure or foundation upon which a model can be further developed and constructed. The archetypes are rarely sufficient models in and of themselves. They are generic in nature and generally fail to reveal important variables that are part of the real system structure of a specific organization. Without an explicit awareness of these real variables, it is difficult for managers to pinpoint specific leverage points where changes in structure can achieve sustainable changes in system behavior.

Number of people: It can be done individually, in groups or just by the decision makers and then presented to other actors in a common meeting.

Time: To share the results and debrief we recommend a 2 h meeting it could be 1 day a week or a month after the other workshop depending on the date of the new meeting to prepare.

Space: We need a safe dialogue space were each actor or group of actors can safely share the subsystems they have identified and those that are missing and the reasons why they believe this is happening.

Equipment: We can use a printed versions of the subsystems identified or a digital one it is important that we can easily add or remove elements from the models.

Set up: We previously ask the actors to try to identify possible subsystems within the SCS and see if they can identify which subsystems are responsible of the 5 main functions (Implementation, Co-ordination, Operational control, Auditing, Development and Policy)

Now we will ask them to try to identify possible pathologies based on the previous workshop.

3.8 Social Transformation Workshop STW#8 "Idealized Design" Interactive Planning Russell Ackoff

STW#8 Idealized Design

Idealized Design and Interactive Planning

Idealized Design: Idealized Design is a methodology developed by Russell L. Ackoff that focuses on envisioning and designing an ideal future state for an organization or system. It involves challenging existing assumptions and constraints to imagine a desired future and then working backward to identify the steps and changes required to achieve that vision.

Key steps in the Idealized Design process typically include:

Identifying the current state, including the strengths, weaknesses, and limitations of the existing system. Imagining an ideal future state without considering the constraints of the present.Analyzing the gaps between the current and ideal states to identify potential innovations and changes.

Developing specific proposals and plans for implementing the necessary changes.

Creating a roadmap for implementing the proposed changes and monitoring progress.

Idealized Design encourages creative and bold thinking, allowing organizations to break free from the constraints of the current state and design innovative solutions and strategies.

Interactive Planning:Interactive Planning is a methodology that emphasizes collaborative and participatory decision-making and planning processes. It recognizes the value of involving stakeholders with diverse perspectives and expertise to develop shared understanding and ownership of the planning process and outcomes.

Ackoff talks about social systems but insists that they are compose by purposeful individuals who can use the system concept of idealized design to think through and pursue the future they desire.

Before idealized design was developed, there were three approaches to organizational planning:

- **Reactivism**—Reactive planners find the solution to their organizational problems in solutions that have worked in the past. They are often nostalgic about the past state of their organizations and speak about "the good old days."
- **Inactivism**—Inactive, or conservative, planners are satisfied with the way things are and hope that their present problems will simply go away if they do nothing. Some observers have compared this mode of thinking to Voltaire's character, Professor Pangloss in Candide, who believed that "this is the best of all possible worlds."
- **Preactivism**—Preactive planners do not look to the past or present for the solution to their problems but believe that the future can be better than the present. For them, the future is an opportunity for improvement to be exploited. The weakness in this approach is in predicting what the future will be. Any prediction of the future ensures a poor outcome. As Yogi Berra wisely observed, "It's tough to make predictions, especially about the future."

  These approaches sometimes worked, but more often they did not. They were especially ill equipped to help organizations adapt to rapid changes in their environment, whether of changes in the market, changes in technology, changes in competitors, or other factors that affect their organizations. Visionary planners began to develop a fourth approach that was to result in the process of idealized design on which this book is based.
- **Interactivism**—Interactive planners reject the approaches of the other three planners. They plan backward from where they want to be to where they are now. They plan not for the future but for what they want their organizations to be at the present time. In so doing, however, interactive managers prepare their organizations for success in the unknowable future.

The Process of Idealized Design

The process of interactive planning, called idealized design, has Two parts:

- Idealization

  1. Formulating the mess
  2. Ends planning

- Realization

  3. Means planning
  4. Resource planning

5. Design of implementation
6. Design of controls

Here is how they work.

Idealization

Formulating the Mess

Every organization or institution is faced with a set of interacting threats and opportunities. These form what we call a mess. The aim of formulating the mess is to determine how the organization would eventually destroy itself if it were to continue doing what it is doing currently—that is, if it were to fail to adapt to a changing internal and external environment, even if it could predict the course of this change perfectly. This process identifies an organization's Achilles' heel—the seeds of its self-destruction—and provides a focus for the planning that follows by identifying what the organization or institution must avoid at all costs.

There are instances in which an organization or institution is faced with a crisis here and now—not sometime in the future. This present mess needs to be understood ("formulated") in the same way as a future mess before an idealized design can be undertaken to avert the possible destruction of the organization. In both cases, the process of formulating the mess is essentially the same.

Formulating a mess involves four steps:

Prepare a systems analysis—a detailed description of how the organization or institution currently operates. This is usually best revealed in a series of flow charts showing how material is acquired and processed though the organization. A similar chart for the flow of money and information is also helpful.

Prepare an obstruction analysis—Identify those characteristics and properties of the organization or institution that obstruct its progress or resist change (for example, conflicts and customs).

Prepare reference projections—Describe what the organization's future would be, assuming no changes in either its current plans, policies, programs, and practices, or changes of what it expects in its environment. This will show how and why the organization or institution would destroy itself unless it makes significant changes. This, of course, is not a forecast but a foresight of how the organization could destroy itself. This projection should reveal how the obstructions described in Step 2 prevent the organization from making adaptive changes to changing conditions.

Prepare a presentation of the mess—Combine the state of the organization and its reference projections into a scenario of the possible future of the organization, a future it would face if it were to make no changes in its current practices, policies, tactics, and strategies, and the environment changed only in expected ways.

Ends Planning
This stage of planning is at the heart of idealized design. It involves determining what planners would like the organization or institution to be now if it could be whatever they wanted. It then identifies the gaps between this idealized design and the organization as it is, thus revealing the gaps to be filled by the rest of the planning process. It is crucial to note here that the design must demonstrably prevent the self-destruction revealed in the formulation of the mess.

Realization

3. Means Planning

This phase requires planners to determine what should be done to approximate the ideal as closely as possible to avoid the self-destruction projected in the formulation of the mess. Planners must invent and select courses of action, practices, projects, programs, and policies to be implemented.

4. Resource Planning

Implementing idealized design requires planners to identify and marshal the resources needed to accomplish the planned changes, including the following:

1. Determine how much of each type of resource—personnel; money; materials and services; facilities and equipment; and information, knowledge, and understanding and wisdom—are required. Also determine when and where to deploy the resources selected.
2. Determine how much of each type of resource will be available at the desired times and places and determine the difference between what will be available in any event and what will be required.
3. Decide what should be done about the shortages or excesses identified in Step 2.

Design of Implementation
Determine who is to do what, when, and where. Create a schedule and allocate resources to the tasks to be carried out.

Design of Controls
Determine:

(1) How to monitor these assignments and schedules,
(2) How to adjust for failures to meet or exceed schedules, and

(3) How to monitor planning decisions to determine whether they are producing expected results (and, if not, determine what is responsible for the errors and correct them).

These six phases of interactive planning do not need to be carried out in the same order presented here, but they are usually begun in this order. Because they are strongly interdependent, they usually take place simultaneously and interactively. Interactive planning is continuous; no phase is ever completed—that is, all parts of a plan are subject to subsequent revision. Plans are treated, at best, as still frames taken from a motion picture.

Constraints and Requirement

There are two constraints imposed on idealized designs and one important requirement. First, the design must be technologically feasible—no science fiction. This constraint does not preclude innovation, but it does restrict innovations to what we currently know we can develop even if we do not have it now. For example, it would be inappropriate in a design of a communication system to use mental telepathy to replace the telephone or e-mail. But clearly, we could increase the functionality of the mobile phone by having it unlock automobiles, turn on their lights, and turn on the heat or air conditioning in the house we are approaching.

The constraint of technological feasibility ensures the possibility of implementation of the design, but it says nothing about its likelihood. An idealized design, however feasible it might be technologically, may not be implementable for economic, social, or political reasons. For example, if all monetary transactions were electronic, a consumption-based tax system—in contrast to an income-based system—would be possible but very unlikely for political reasons. The second constraint is that the design, if implemented, must be capable of surviving in the current environment. Therefore, it cannot violate the law and must conform to any relevant regulations and rules. It does not mean that the design must be capable of being implemented now. It does mean that if the design were implemented now, it would be able to survive in the current environment. For example, it would be possible to implement a system of all-electronic voting in elections, but it would not survive in today's world of computer hacking where voters cannot be sure that their votes are being counted. In the future, however, when voters can be confident of the integrity of the system, it will probably be implemented.

Finally, there is the important requirement that the process that is designed must be capable of being improved over time. If that which is designed is an organization or institution, it must be capable of learning and adapting to changing internal and external conditions. It should be designed to be ready, willing, and able to change itself or be changed. Therefore, the product of an idealized design is neither perfect, ideal, nor utopian, precisely because it can be improved. However, it is the best ideal-seeking system its designers can imagine now.

Anticipating The Future

We have pointed out how difficult it is to predict the future. And idealized design stresses the need for planners to concern themselves with what they want now, not at some future time. However, this does not remove the need to take the future into account. It changes the way the account of it should be taken. In conventional planning, to exist. Unfortunately, as the rate of change in the environment continually increases, along with its complexity, accurate forecasting becomes more and more difficult and less and less likely. As we have observed, poor forecasts (or predictions) lead to poor outcomes. How then should the future be taken into account? The future is taken into account in idealized design by the assumptions planners make about it. Contrary to what some forecasters claim, assumptions about the future differ qualitatively from forecasts. Forecasts are about probable futures; assumptions are about possible futures. We carry spare tires in our cars even though we do not forecast having a flat tire on our next trip. In fact, if anything, we forecast that we will not have a flat tire on the next trip. But we assume a flat tire is possible, however unlikely it may be. Assumed futures can be taken care of in two different ways. First, there is contingency planning. When there are a relatively few and explicitly describable possible futures, planners can prepare plans for each possibility. This is called contingency planning. Then, when the truth about the future is known, the appropriate plan can be invoked. For example, an oil company can develop exploration plans based on the price of oil increasing, staying the same, or declining. When it is apparent how the price is moving, they can quickly move to the appropriate plan already developed. The way of dealing with more contingencies than can be planned for separately is to design into the organization or institution enough flexibility and responsiveness so that it can change rapidly and effectively to meet whatever it encounters. Automobile manufacturers cannot accurately predict customer demand for all possible models, colors, and accessory packages. However, the best automakers have solved this problem by designing production lines that allow them to build different models and colors on the same production line as customer demand requires. Some manufacturers in several industries have created such flexible production facilities that they can customize each individual product based on an order just received. Boeing aircraft and Dell computers are examples. It is obvious that an additional benefit of such a system is that it allows for a rapid inventory turn and minimum idle capital.

Effects of Idealized Design

So far, we have described the way idealized design is put into practice. However, planners should be aware of an additional dimension to the process. It has a number of beneficial effects on those who engage in it and on their organizations, as follows:

- Promotes understanding of that which is designed
- Transforms the designers' concept of what is feasible
- Simplifies the planning process

- Enhances creativity
- Facilitates implementation

Let's look at each in turn.

Promotes Understanding

There is no better way to gain an understanding of something than by designing it. Designing something as simple as a door handle on a car requires the designer to understand how the human hand grasps a handle and then turns (or pulls) so that the design produces a comfortable and functional handle. Furthermore, in the design process, for example, one is forced to consider the assumptions on which the design is based. This consideration frequently reveals the irrationality of some of the features of the existing object and allows for their replacement. For example, in nearly all men's stores, clothing is arranged by type; a section for suits, another for overcoats, another for shirts, and so on. When a group of male planners engaged in an idealized redesign of a men's store, it became apparent to them that this arrangement was for the convenience of those who run the store, not its customers. They found that a far better arrangement for customers was to arrange the garments by size, not type of clothing, putting all the suits, coats, shirts, and so on in the same place so that each shopper—small, medium, or large—could find everything he might want in one place. Bookstores have always known this and arrange books by subject (because most browsers know what interests them, even if they do not know which books are available).

Transforms Designers' Concept of Feasibility

The principal obstruction to what we want most is ourselves. The great American philosopher Pogo recognized this in his classic observation that "We have met the enemy and he is us." Our tendency, however, when we stand where we are and look toward what we want, is to see all kinds of obstructions imposed from without. When we change our point of view and look backward at where we are from where we want to be, in many cases the obstructions disappear. Banking is a good example. Years ago, banks employed many tellers who handled transactions with customers. They received deposits and filled out deposit slips, cashed checks, and entered interest in savings passbooks. Bankers had to hire legions of tellers as their business grew. However, a few visionary bankers asked themselves what would be the ideal bank. They concluded that it would have few—perhaps no—tellers and would process all the same transactions. This vision led them to create automatic teller machines that allowed customers to do the work rather than the tellers. In turn, this led to online banking, where customers do not even have to go to the bank to manage their accounts. The obstruction bankers thought they faced—how to find and pay all those tellers—disappeared when they realized that banks could operate just as well with a decreasing number of tellers. Although some customers complained about this change, many more were pleased at not having to stand in line waiting to be helped by a human being.

Simplifies The Planning Process

Planning backward from where one wants to be reduced the number of alternatives that must be considered when making a choice of how to get there. It simplifies the planning process considerably. An organizational example of simplification—requiring the details of planning backward and forward—is too long for our purposes here. So we offer instead an example drawn from a tennis tournament that nicely encapsulates how working backward greatly simplifies idealized design. If 64 players enter a tennis tournament, how many matches must be played to determine the winner? This is not hard to determine. There will be 32 matches in the first round, then 16, 8, 4, 2, and 1, successively. Added together, these equal 63 matches. However, if we start at the end and ask, "How many losers would there have to be?" the answer is obviously 63, and no arithmetic is required. The advantage of working backward is even more apparent if we start with several players that is not a power of 2, say 57. The arithmetic now becomes complicated because some players must be exempted from the first round to make the number of players left after that round a power of 2. If we work backward, however, it is apparent that there must be 56 losers: hence this number of matches.

Enhances Creativity

Human creativity is as old as humankind, but it was not very long ago that we began to understand what it is. We believe that it is a three-step process. First, it requires that we identify a self-imposed constraint, an assumption that we make consciously or unconsciously that limits the number of alternatives we consider. Second, we must deny or eliminate that assumption as too limiting. Third, we must then explore the consequences of this denial. These steps are conspicuous in solving a puzzle (because a puzzle is a problem we cannot solve if we make an incorrect assumption). When the solution to a puzzle we have not been able to solve is revealed to us, we want to kick ourselves because we realize that we were the obstruction between the puzzle and its solution.

Facilitates Implementation

A major reason most plans are not fully implemented is that those people responsible for implementing it have no sense of ownership of it. This leads to resentment and subversion of its implementation. Idealized design, however, requires the participation of everyone who will be affected by it. Therefore, ownership of the resulting plan is widely spread among those who must implement it. This avoids resistance and subversion. Implementation of a design and plan based on it is usually carried out enthusiastically by those who had a hand in preparing it.

Interactive planning methodology:

Interactive Planning is of particular relevance to us in this book because it was specifically designed to cope with the "messes" that arise from the increased complexity of the modern era. Three principles underpin the methodology (Ackoff 1981 1999)

The first is the participative principle. If possible, all stakeholders should participate in the various phases of the planning process. This is the only way of ensuring "objectivity". It also secures the main benefit of planning- the involvement of members of the organization in the process. This is more important than the actual plan produced. It is by being involved in the process that stake holders come to understand the role they can play in the organization. It follows, of course, that one can plan for anyone else. The role of professional planners is not to do the planning, but to help others plan for themselves.

The second principle is that of continuity. Because values change and unexpected events occur, plans need to be constantly revised.

The third is the holistic principle. Because of the importance of the interactions between the parts of a system, we should plan simultaneously for as many parts and levels of the organization as possible.

With these principles in mind we can now consider the five phases of the Interactive Planning approach:

1. Formulating the mess
2. End Planning
3. Means Planning
4. Resource Planning
5. Design of implementation and control.

These phases may be started in any order and none of them, let alone the whole set, can ever be regarded as completed. They constitute a systemic process.

Formulating the "mess" involves determining the future an organization will be in if it continues its current plans, policies, and practices and if its environment changes only in ways that are expected. Three types of study are necessary:

1. Systems Analysis: Giving a detailed picture of the organization, what it does, its stakeholders and relationships with its environment?
2. Obstruction and Analysis: Setting out any obstacles to corporate development.
3. Reference projections. Extrapolating on the organization´s present performance in order to predict future performance if nothing is done and trends in the environment continue in entirely predictable ways.

Interactive Planning was developed by Russell Ackoff to assist stakeholders design a desirable future or idealized design for themselves, their organizations, and their organizations environments and to help them find ways of bringing that future about that Idealized design. This is a powerful vision of the role of management science and explains in large part why Ackoff´s work has had such a huge impact on the OR and systems communities. Examined in detail, and in terms of what it can deliver, interactive planning does not disappoint. It is arguable, however, that no approach can be comprehensive enough to achieve what Ackoff wants from interactive planning and that, in representing itself as being comprehensive, it inevitably disguises its own biases (Figs. 3.11 and 3.12).

About this workshop

Our society and the principle private and public organizations that it contains have reached a level of maturity that eliminates whatever effective applies deterministic and animalistic models to social systems may once have had.

Interactive planning was specifically designed to cope whit messes that arise from the increased complexity of the modern era.

5 steps of Interactive planning

1. Formulating the mess
2. Ends planning
3. Means planning
4. Resources planning
5. Design of implementation and control
6. Past present Future

Number of people: It can be done individually, in groups or just by the decision makers and then presented to other actors in a common meeting.

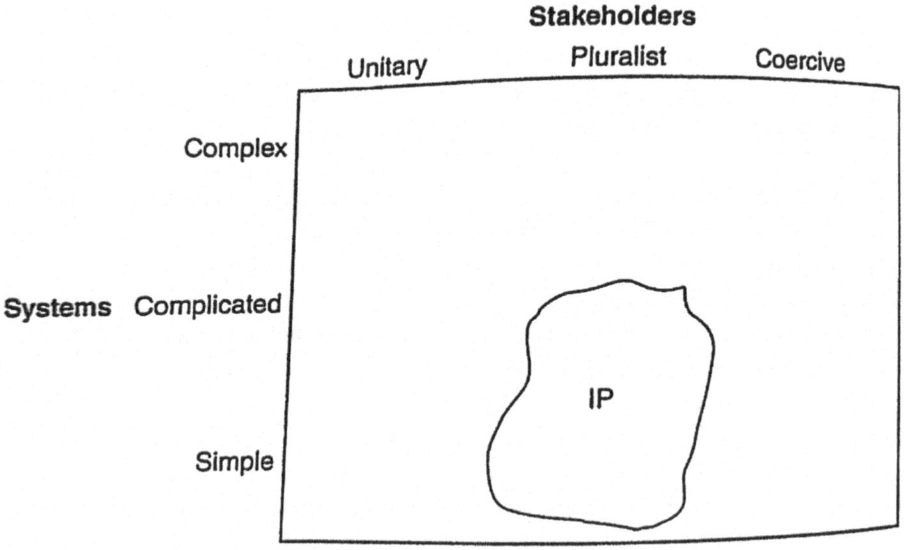

Fig. 3.11 The positioning of interactive planning on the SOSM

ORIENTACIÓN	PASADO	PRESENTE	FUTURO
Reactivista	+	-	-
Inactivista	-	+	-
Preactivista	-	-	+
Interactivista	+/-	+/-	+/-

Fig. 3.12 Different types of planning 2020

Time: To share the results and debrief we recommend 2 h meeting it could be 1 day a week or a month after the other workshop depending on the date of the new meeting to prepare.

Space: We need a safe dialogue space were each actor or group of actors can safely share the subsystems they have identified and those that are missing and the reasons why they believe this is happening

Equipment: We can use a printed versions of the subsystems identified or a digital one it is important that we can easily add or remove elements from the models

Set up: We previously ask the actors to try to identify possible subsystems within the SCS and see if they can identify which subsystems are responsible of the 5 main functions (Implementation, Co-ordination, and Operational control, Auditing, Development and Policy)

Now we will ask them to try to identify possible pathologies based on the previous workshop.

3.9 Social Transformation Workshop STW#9 "Conceptual Model" Peter Checkland Systems of Systems Methodology

STW#9 Conceptual Model

> Peter Checkland is known for developing the Soft Systems Methodology (SSM), which is a problem-solving and systems thinking approach. Within SSM, the use of conceptual models plays a significant role in understanding complex problem situations and facilitating communication and learning.
>
> (Checkland 1981).

Conceptual Models: Develop conceptual models that describe different perspectives or viewpoints of the problem situation. These models can take various forms, such as system diagrams, flowcharts, or influence diagrams, and represent the structure, processes, and interactions within the system.

About this workshop

At its simplest, The Conceptual Model workshop allow us to enable a better, more focused debate to take place. This in turn, made participants more willing to engage with the pre-prepared models the facilitators would sometimes introduce to further enrich and bring more rigor to the ongoing discussions (Fig. 3.13).

To build a conceptual model from a particular worldview allows us to appreciate the roll this could play enabling a better, more focused debate to take place.

For the applied scientist, scientific method must include a philosophy... It is what Germans call a Weltanschauung, a worldview, a perception of what reality is like... The

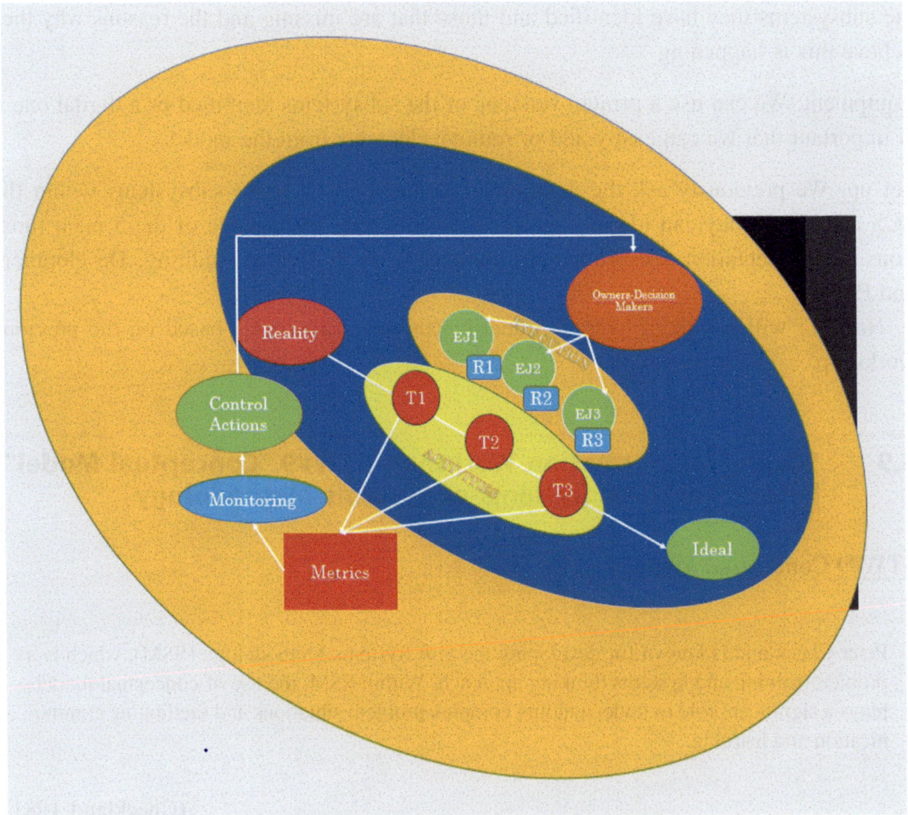

Fig. 3.13 Conceptual model Navarro 2018 based on SSM checkland Peter

applied scientist is not merely applying the results of pure research; he is applying his Weltanschauung.

> "*The experience, especially the experience of human communication, develops in each of us "readiness" to notice particular aspects of our situation, to discriminate them in particular ways and to measure them against particular standards of comparison, which have been built up in similar ways.*"
>
> Jackson (1972).
>
> "*An organization is an abstraction: It is a social collectivity concerned with some collective action, and there are associated social practices which relate to this*".
>
> "*Members of organizations are not necessarily simply contributors to the achievement of organizational goals, as the conventional model suggest.*"
>
> Checkland and Holwell (1992).
>
> "*to incorporate the concept of worldview it is necessary to abandon the idea that the world is a set off system instead the world is taken to be very complex, problematical, mysterious, characterized by clashes of worldview. It is continually being created and recreated by people thinking talking and taking action.*"
>
> Checkland and Poulter (2006).
>
> Assumptions:
>
> 1. Social Reality is socially constructed continuously.
> 2. Use explicit intellectual devices consciously to explore, understand, and act in the situation in question.
> 3. Include in the intellectual devices "Holons" in the form of systems models of purposeful activity deal on the basis of declared worldviews.
>
> Checkland and Scholls (1990).

To Run this Workshop:

Number of people: It can be done individually, in groups or just by the decision makers and then presented to other actors in a common meeting.

Time: To share the results and debrief we recommend 2 h meeting it could be 1 day a week or a month after the other workshop depending on the date of the new meeting to prepare.

Space: We need a safe dialogue space were each actor or group of actors can safely share the conceptual model they develop.

Equipment: We can use a drawing version of the Conceptual model identified or a digital one it is important that we can easily add or remove elements from the models.

Set up: We previously ask the actors to create their own conceptual model to share with others and create a common one.

3.10 Social Transformation Workshop STW#10 "Gaps and Priorization"

STW#10 Gaps and Priorization

About this workshop

In the Idealized Design methodology, identifying gaps and prioritization are important steps in the process of transitioning from the current state to the envisioned ideal future state. These steps help ensure that efforts are focused on the most critical areas and opportunities for improvement. Here's an overview of how gaps and prioritization are addressed in the Idealized Design methodology:

Identifying Gaps: Once the ideal future state has been envisioned, the next step is to analyze the gaps between the current state and the desired future state. This involves identifying the discrepancies, shortcomings, or deficiencies that exist in the current system that prevent it from aligning with the envisioned ideal state.

Gaps can be identified through various means, including:

Conducting a thorough analysis of the current state, assessing strengths, weaknesses, and limitations.

Comparing the current system against the idealized vision and identifying areas where the current state falls short.

Engaging stakeholders, gathering their perspectives, insights, and feedback on what is lacking or needs improvement in the current system.The gaps identified serve as a starting point for determining the areas that require attention and change in order to move closer to the ideal future state.

Prioritization:After identifying the gaps, the next step is to prioritize the identified gaps based on their significance and potential impact on achieving the ideal future state. Prioritization helps allocate resources and efforts effectively, focusing on the most critical areas for improvement.

Priorization can be done through various means, including:

Assessing the potential benefits and impact of addressing each identified gap.

Consider factors such as the alignment with the overall vision, the magnitude of the gap, and the feasibility of addressing it.

Engaging stakeholders to gather their perspectives on the relative importance of different gaps and the potential consequences of addressing or not addressing them.
Considering any dependencies or interdependencies between gaps and the potential ripple effects of addressing one gap on others.
Through prioritization, attention is directed toward the most significant gaps, ensuring that efforts and resources are concentrated on areas that will have the greatest impact on achieving the idealized future state.
By systematically identifying gaps and prioritizing them, the Idealized Design methodology helps guide the process of designing and implementing changes to bridge the identified gaps and move closer to the desired future state.
At its simplest, the gaps and priorization workshop allow us to reflect on the impact and effort and the importance and urgency of the different actions we can take to reduce the identified gaps.

This way we can give priority to the most important activities and focus our time and resources based on the available information and what we can do with it.

Contrasting 'is' and 'ought' boundary judgments provides a systematic way to evaluate the normative content of planning while at the same time laying open the normative basis of the evaluation itself: The 'is' questions aim at determining a design's effective normative implications in the light of the 'ought' answers, that is to say, without any illusion of objectivity. Understanding our reality and its characteristics as we did with the interactive planning methodology allow us to evaluate the current state of our studied system. Identification and prioritization of possible action gaps has the potential to lead to more rapid generation of subsequent successful projects informed by input from stakeholders Audiences including researchers, funders, clinicians, advocates, and patients could use information about prioritized research gaps to understand areas of uncertainty and more quickly initiate studies.

To Run this Workshop:

Number of people: It can be done individually, in groups or just by the decision makers and then presented to other actors in a common meeting.

Time: To share the results and debrief we recommend 2 h meeting it could be 1 day a week or a month after the other workshop depending on the date of the new meeting to prepare.

Space: We need a safe dialogue space were each actor or group of actors can safely share their priorization model they develop.

Equipment: We can use a drawing version of the Priorization model identified or a digital one it is important that we can easily add or remove elements from the models.

Set up: We previously ask the actors to create their own Priorization model to share with others and create a common one.

We will establish the most important gaps using criteria.

- Importance:
- Urgency
- Impact
- Effort
- Incremental Value
- Feasibility.

We had previously defined the characteristics of our reality and the idealized system, now we want to identify gaps that exist between the 2 systems: Once the Gaps have been identified, we think about possible actions (Transformations to close these gaps).

Model Is - Reality	Model Should Be - Ideal	Gap	Necessary transformation to close that Gap

For the priorization we ask them to consider several criterion as Current importance, potentiality for impact, incremental value and feasibility.

More important Gaps:

Proposed process to close that gap.

3.11 Social Transformation Workshops STW#11 "Design or Re-design a Project"

STW#11 Design or re-design Projects

About this workshop

At its simplest, the design and redesign a project workshop seeks to establish the precise actions to successfully plan, execute, monitor, and control the implementation of a project assuring continuity and legitimacy of this endeavor by actively involving different actors of SCS in a participatory way.

Projects are "a temporary endeavor undertaken to create a unique product, service or result" (PMI 2008)

Projects are carried out to meet objectives by production of deliverables.

An objective is defined as a goal towards which work should be directed, a strategic position to be achieved, an end to be achieved, a result to be obtained, a product to produce or a service to provide.

A deliverable is defined as any single product, result, or capacity and verifiable to run a service that is produced to complete a process, phase, or project. The Deliverables can be tangible or intangible.

Projects operate within the constraints imposed by the organization through its structure and framework of governance. To operate effectively and efficiently, the project manager needs to understand where they reside. Responsibility, accountability, and authority within the organization. This knowledge will help the project manager to effectively use his power, influence, competence, leadership, and political capacities to successfully complete the project. The interaction of multiple factors within an individual organization creates a unique system that influences the project that operates in that system. The resulting organizational system determines power, influence, interests, the competence, and political capacities of people who can act within the system.

Projects drive change. Projects drive change in organizations. From a business perspective, a project is intended to move an organization from one state to another state in order to achieve a specific objective.

Before the project starts, it is usually said that the organization is in the current state. The desired outcome of the project-driven change is outlined as the future state. For some projects this may involve creating a state of transition, where they are carried out multiple steps along a continuum to reach the future state.

The successful completion of a project it leads the organization to move into the future state and achieve the specific objective.

Projects drive change. Projects drive change in organizations. From a business perspective, a project is intended to move an organization from one state to another state in order to achieve a specific objective.

The temporary nature of projects implies that a project has a principle and a defined end. Just because it's temporary doesn't necessarily mean a project is short-lived. The end of project is reached when one or more of the following situations are met.

The project manager plays a critical role in the leadership of a project team in order to achieve the objectives of the project. This role is clearly visible throughout the project. Many project managers get involved in a project from its initiation to its closure. However, in some organizations, a project manager may be involved in evaluation and analysis activities prior to project initiation. This activities may include consultations with business unit and executive leaders about ideas to advance goals strategic, improve the performance of the organization or meet the needs of customers. In some contexts, organizational, the project manager may also be required to direct or assist in business analysis, the development of the business case and aspects of portfolio management for a project. A project manager You may also be involved in follow-up activities related to obtaining business benefits of the project. The role of a project manager may vary from one organization to another. Ultimately, the role of Project management is adapted to fit the organization, in the same way that the processes of project management Projects are tailored to fit the project.

(PMBoK 2017)

To Run this Workshop:

Number of people: It can be done individually, in groups or just by the decision makers and then presented to other actors in a common meeting.

Time: To share the results and debrief we recommend 2 h meeting it could be 1 day a week or a month after the other workshop depending on the date of the new meeting to prepare.

Space: We need a safe dialogue space were each actor or group of actors can safely share their priorization model they develop.

Equipment: We need the results of the gaps & priorization workshop to be able to create a process of design or redesign to plan execute, monitor & control the project.

Set up: We previously ask the actors to create their own Priorization model, based on this priorization model we will create a step by step action project.

3.12 STW#12 "Systems Dynamics"

STW#12 System
System Dynamics

> System dynamics is an interdisciplinary approach to understanding and modeling the behavior of complex systems over time. It was developed by Jay W. Forrester and has been widely applied in various fields, including management, engineering, economics, and environmental studies.System dynamics focuses on the dynamic interactions and feedback loops within a system and how they lead to the system's behavior over time. It recognizes that systems are composed of interconnected components that influence each other, and changes in one part can have non-linear effects on the entire system.
>
> "*Systems thinking is a discipline for seeing the structures that underlie complex situations, and for discerning high from low leverage change…Ultimately, it simplifies life by helping us to see the deeper patterns lying beneath the events and the details.*"
>
> (Senge 1990)

In 1968 a group of prominent international politicians, government officials, diplomats, scientists, economists, and business leaders met in Rome to discuss the present and future "predicament of mankind". Their conviction was that the problems facing mankind were so complex and interrelated that they were beyond the scope of traditional institutions and individual policies and needed to be studied and addressed "*as a whole*". This led them to find a multidisciplinary think-thank that became known as the "Club of Rome" One of the Club´s earliest initiatives was to commission an international team of researchers, centered on the Massachusetts Institute of Technology (MIT) and funded by the Volkswagen

Foundation, to produce a model that described the behavior of the different but interrelated components that make up *"the global system"*. To achieve this, the 17 researchers employed the global model and system dynamics (SD) approach set out by Jay Forrester in his book World Dynamics (1971) their report, known as the limits to Growth (1972), sent shock waves around the world. Its foremost conclusion stated starkly that:

> *"If the present growth trends in world population, industrialization, pollution, food production, and resource depletion continue unchanged, the limits to growth on this planet will be rather sudden and uncontrollable decline I both population and industrial capacity."*
>
> (Meadows et al. 1972).

To many readers and commentators, it seemed that doomsday was just around the corner.

The researchers took the five key factors influencing the behavior of the global system to be population growth, industrialization, food production, consumption of nonrenewable natural resources, and pollution. Looking at the statistics from 1900 onward they all seemed to be growing exponentially. But the world was finite. If the growth of population and industrialization continued at their current rates, then a point would come when there was no new arable land to provide increased supplies of food, nonrenewable resources became severely depleted, and pollution began to endanger the natural environment through the production of CO2, global warning, and radioactive wastes. But what were the exact limits to growth and when would they bite? To find out the researchers elaborated on their intuitive conclusions using a computer model, WORLD 3 based on the principles of System dynamics set out by Forrester. Experts were consulted for their opinion on the interrelationships between the five elements. They pointed around 100 casual links and helped quantify the force of the interactions. Positive and negative feedback loops and lags were identified. Data about the five components could now be plugged in and the model run to produce an output that showed the effect of the simultaneous operation of all the relationships. The model made it possible to identify the critical determinants of behavior and to look at what happened if the numerical values of the initial assumptions were changed. Different policy proposals could also be tried out and their effects were monitored over time.

The standard run of the model based on there being no significant change in the trends showed that well before 2100 resource depletion would lead to a collapse of economic development and a decrease in the global population. Changing the assumptions on which the model was built could extend the period before collapse but, however drastic those changes were there was always collapse before 2100. If not due to resource depletion, then because of food shortage or pollution. For example, if resource depletion and growth in pollution were reduced to a quarter of their current levels, starting in 1975, then collapse still occurred because of food shortage. Even if it was assumed that there were significant technological innovations, unlimited resources, and more effective birth control initiatives, the system collapsed due to pollution. The fact that significant changes in input data did not alter the overall pattern of behavior convinced the researchers that system dynamics was correct to assume that the feedback structure of the model was a more important determinant of behavior than the actual numbers that are used to quantify the feedback loops. Thus, while they were clear that the model could not produce exact predictions, they were able to insist that:

> "*The basic behavior mode of the world system is exponential growth of population and capital, followed by collapse.*"
>
> (Meadows et al. 1972).

All was not lost. The model did not say that the future was determined. Social factors were excluded from the model because they were much too complex to deal with. The authors of the 1972 report now reintroduce them into their thinking and argue that new forms of social behavior are needed to avoid catastrophe. If we can learn from the simulation and take action to weaken the positive feedback loops, by imposing our own limits to growth of population and industrial output, then a sustainable future is possible in which the basic material needs of all people are met. We must start immediately to make an ordered transition from growth to global equilibrium. This was going to require a huge shift in human values and the Executive Committee of the Club of Rome end their commentary on the report with:

> "*The last thought we wish to offer is that man must explore himself–his goals and values as much as the world he seeks to change. The dedication to both tasks must be unending.*"
>
> (Meadows et al. 1972).

The publication of limits of Growth provoked uproar and brought forth a tide of criticism. Much of this related to the authors giving too little weight to possible scientific and technological advances. They had, in fact, allowed for these but calculated that they would come too late to prevent collapse. Some of the criticism had merit, as when the Science Policy Research Unit, at the University of Sussex, suggested that the simulations were more sensitive to a few key assumptions than the researchers recognized (SPRU 1973). Some of it was sensible, asking how developing countries would react to calls for a curb on growth. Some of it, however, was blatantly ideological. Whatever the facts, the Catholic Church had to resist arguments in favor of birth control; politicians has to champion growth and classical economists had to believe in equilibrium models in which prices would always ensure that a correction took place. The report's conclusions were dismissed, even ridiculed.

By the early years of the twenty first century, however, attitudes to The limits to Growth began to shift drastically. Follow-up research by Donnella Meadows (2004) and other independent investigators, based on data from 1970 to 2000 showed that the results of the standard run of the model gave good approximations of what had happened for almost all the reported outputs. The global system was on track to collapse sometime in the second half of the century. A recent report of UK All Party Parliamentary Group reviewing the latest research concludes that

> "*...there is unsettling evidence that society is still following the standard run of the original study in which overshoot leads to an eventual collapse of production and living standards.*"
>
> (Jackson and Webster 2016).

Sixty years ago, Jay Forrester (2013) MIT professor, generalized those boxes and arrows to represent the stimulating or inhibiting influences in any system, first applying this ***"system dynamics"*** to the economics of business cycles.

The reification of these positive and negative influences from one factor to another as boxes representing levels or circles representing variables and the directed arrows between them are known as "***Causal Loop Diagrams***."

Other Authors as Linda Sweeny and Dennis Meadows (2010) developed games and exercises for such "systems thinking" for K-12 teachers. Joseph Novak (2009) developed causal loop diagrams into "**Concept Maps**" of abstract concepts as a way for students to organize and understand the material they are trying to learn.

A causal loop diagram pictures the named individual elements of a system as the **"nodes"** of a graph and the links between them as the **"edges"** of that graph. The links are causally directed from one node to another. The strength of the links may all be the same, or they can differ from each other. The links may also either be **"positive"** meaning that an increase in the value of the originating element increases the value in the target element, or **"negative"** meaning that an increase in the initiating element decreases the value in the target element. Typically, positive links are identified by a **"+"** sign or a color (such as blue) and the negative links by a **"–"** sign or a color (such as red).

This dynamic is a consequence, fundamentally, of the interactions between the elements of the system. The interactions between the elements are setting up, on the other hand, the structure of the system. That is why, from a point of view Dynamic view, every system is being determined by its structure. And in this one In this sense, one of the most outstanding aspects of System Dynamics is that the causes of problems that appear in a social system are not so due to previous events as well as to the structure of the system itself (Forrester 1961, 1968; Aracil 1983; Martínez and Requena 1986; Senge 1990).

Interested in the dynamics of systems, especially those of an industrial type, Jay W. Forrester, an MIT engineer, will develop a whole new discipline with a marked technological character. We could define the Dynamics of Systems as a way of studying the behavior of systems, of a social type, to show how policies, decisions, structures and delays in feedback flows interrelate to influence the development and stability of own systems (Forrester 1961). More specifically, we can conceive of the Dynamics Systems as a methodology for the construction of models of social systems (Aracil 1983). It is intended to establish techniques to formalize, in a mathematical way, the models that verbally or mental settle on these systems (Navarro 2001).

The origin of System Dynamics is found in a demand for an electric company for the study of the oscillations presented by their orders (Forrester 1961). In this initial study, Forrester already realized the importance of two factors for understanding the dynamics of a system: the type of feedback loops that were taking place in it and delays or delays that were produced in these own feedback loops. Working.

With these two elements, Forrester was able to give a satisfactory answer to the problem of oscillations in orders. Since its inception, Forrester has strongly linked Systems Dynamics to business management. So much so, that he conceived of industrial dynamics-that was the original name he gave it-as a discipline that would help managers in their everyday decision making. 'The manager,' Forrester says, 'may have a laboratory design to help create improved control policies and information flows on which they depend. This will lead to a greater depth in management education and changes in the capabilities of managers' (1961 Spanish edition). From here, Forrester continued working on the modeling of different systems, among whose most prominent examples stand out one on urban dynamics (1969) and an overly globalizing attempt on a dynamic of the world (1971).

In the genesis and later development of Systems Dynamics contributed disciplines such as the theory of servomechanisms and general systems theory. The theory of servomechanisms focuses on the systematic study of the concept of feedback and dynamic behavior of systems The theory of the servomechanisms is therefore a germ that would later be developed under the name of cybernetics by different authors, especially Norbert Wiener. To all this must be added the strong evolution of computers late 1950s, allowing easy modeling of simulations.

The use of the computer in the simulation of the created models is basic. The computer allows to show dynamic consequences of the interrelations between the components of the system that escape mental processing by their strong tendency to think in linear and close cause-effect relationships over time and space (Aracil 1983; Senge 1990). The simulation in the computer allows us to penetrate the complexity of social systems (Wagensberg 1985). It allows working with non-linear relationships, a type of relationship that does not usually fit well in our usual way of thinking. The simulation on the computer, replaces in this way to laboratory experimentation, experimentation that, for.

On the other hand, it is difficult to execute when we are interested in social systems. Therefore, 'simulation can play the role of experience versus theory' (Wagensberg 1985) as long as you can put in question it and suggest its revision and, furthermore, 'simulation can play the role of theory versus experience) in in that it can describe, understand and predict phenomena without the need for put them to the test.

Basically, and very schematically, a study in system dynamics presents the following phases (Aracil 1983):

1. Observation of the behavior of the system to be able to identify fundamental elements that compose it and that are of interest to the study in particular of it.
2. Determination of the feedback structures that can produce the observed behavior.
3. From the identified structure, the mathematical model is built behavior of the system so that it can be simulated in a computer.
4. Simulation of the model.
5. The proposed structure is modified until its components and behavior result match the behavior observed in the real system.
6. Simulation of decisions on the computer: decisions are modified that can be introduced into the simulation model until acceptable and usable decisions are found that result in real behavior improved.

Systems Dynamics involves the detailed study of the interrelationships of the system for its subsequent modeling in order to be able to simulations. Simulations that will be of great help for the subsequent intervention in the system itself. Simulate before intervening.

In this way, the usefulness of System Dynamics in terms of the realization modeling and subsequent simulations are an invaluable aid for the management decision-making, in both private and public spheres, as have revealed numerous works (Merten 1988; Richardson and Rohrbaugh 1990; Senge and Sterman 1992; Aitchison and Kocay 1994; Vennix 1995; What my, Larse and Ginsberg 1997). And, furthermore, also in this sense there have been authors who have been able to pick up one of Forrester's initial concerns and have linked the Systems modeling and its simulation with management education (Senge 1990; Graham and Senge 1990; Carena and Castelli 1993; Sterman 1994; Eftekhar Strong and Hawalenskha 1997). Systems Dynamics as a useful tool in the learning a systemic thinking, basic for the management of social systems. For those in favor of this procedure, the use of simulations supposes a decision-making method clearly superior to methods based on simple accumulation of experiences. Experience is shown to constitute a very primary learning and does not obtain the most relevant information in the functioning social systems (Senge 1990; Senge and Sterman 1992; Sterman 1994; Kim and Senge 1994; Lomi, Larsen and Ginsberg 1997). (Navarro 2001).

And this is basically due to three reasons:

1. First, the relationships between elements tend to be not linear, which makes the social system more complex.
2. Second, the existence of defaults or the distance in time between causes and effects and, third, how difficult it is to human thinking results from understanding feedback loops, even more so when there are several loops present.

3. Ultimately, it's about limitations humans in the cognitive processing of information. This being the case, we could expect high evidence about the greater efficacy of decision-making guided by simulation of systems that decision-making not based on it. In a study by Cavaleri and Sterman (1997), and the only that we have found in this regard, working with several agencies insurance firms in the US that began to apply simulation based on the Systems Dynamics as an aid for decision-making, find evidence average having the productivity of the agencies as a criterion variable of success. In some cases, productivity increased, in others it did not increase at all. Cavaleri and Sterman emphasize the importance of correct systems modeling. The key aspect for the modeling of systems to be of help in the taking of decisions is in the consideration of the important aspects of the concrete phenomenon and in the correct determination of the relationships between variables that constitute said phenomenon.

Central elements in System Dynamics.

Causal Loop Diagrams

People tend to think in terms of unidirectional cause-effect relationships, forgetting other types of more complex relationships as we have shown numerous studies and experiments (e.g. Hall 1976, 1989; Kleinmuntz 1985; Sterman 1989a, b; Brehmer 1989). Hence the concept of feed-back or feedback of cybernetics has been a small revolution that has shaken our most classical conception of causality. In essence, the concept of feedback comes to emphasize that separating cause from effect is a mere analytical device that, moreover, constrains thought insofar as obvious part of the complexity of the relationship. An example. In an act as simple and everyday as filling a glass of water, we usually express ourselves with words like "I make the water level rise as far as I want". In more causal language: I cause the level to rise of water (effect). Cybernetics emphasizes that such a fact is more complex, something like the following:

> My purpose of filling a glass of water creates a system that causes water to flow when the level is low and then turn off the tap when the glass is full
>
> (Senge 1990).

Focusing on the importance of feedback loops or curls, let's distinguish between negative feedback loops and positive feedback loops, classical distinction within System Dynamics and Cybernetics (Forrester 1961, 1968; Maruyama 1963; Aracil 1983; Martínez and Requena 1986; Briggs and Peat 1989, Senge 1990).

A well-known example of a negative feedback loop is found on the thermostat. We program the heating in our living room so that the temperature is kept at 18 ºC and the thermostat contained in this appliance makes that the stove works more or less intensely to reach this value. When the temperature drops below 18º the thermostat turns on the stove to heat the living room and, when the temperature exceeds 18º, the thermostat turns off the stove. The action of the thermostat affects the stove, but also the action of the stove affects the thermostat.

In dynamic terms we would say that a negative feedback loop causes the variation in an element to be transmitted along the loop in such a way that determines a variation of the opposite sign in that same element, that is, any variation that occurs in one of the elements of the loop tends to cancel itself out. Otherwise, negative feedback loops tend to stabilize the system, to keep it in a value, series of values or a certain state. In a goal or objective. For this reason, they are compensatory feedback mechanisms, containment of deviations.

The interest in negative feedback loops was found in the middle of the nineteenth century in the theory of servomechanisms, the germ of which mid-40 s Norbert Wiener would baptize as Cybernetics. Wiener worked on the improvement of the antiaircraft artillery, which did not quite work due to at the speed of airplanes. When going from shooting at targets fixed to moving targets, Wiener worked on the prediction of the mobile's flight, therefore,

It was "*of utmost importance to fire the missile, not at the target, but so that the missile and target coincide at a certain future moment in space" (Wiener 1948). To this end, Wiener used the information provided by the difference between the prediction made at each moment and the actual position of the white: "when we want a movement to follow a certain pattern, the difference between this pattern and the movement actually performed is used as new momentum for the regulated part to move in such a way that its movement is get closer to the one foreseen by the employer*".

We see that our usual cause-effect description contemplates the part right of the diagram, from "tap position" to "current water level". Description which, we now verify, is partial.

We see that this more classical cybernetics has a marked teleological character, due to the character of negative feedback loops. In the case before described, to hit the target.

But not all feedback loops are negative. Also exist feedback loops in which, instead of dampening the effects, the effects are amplify and, which have served Maruyama (1963) to speak of second cybernetics. Positive feedback loops act in such a way that the variation of an initial element spreads along the loop reinforcing it. Coupling a sound serves as an example. The audio output from the speaker is captured by the microphone that reintroduces it, the amplified sound comes out again that re-enters with greater intensity through the micro, and in a few tenths of a second the annoying sound that is well known to us is produced. Positive feedback loops represent a "snowball effect", a circle virtuous.

In dynamic terms, a positive feedback loop represents the variation of an element that propagates through the loop and reinforces the variation initial making her older. Therefore, positive feedback loops they tend not to maintain equilibrium, on the contrary, they transport the system towards new state. They are reinforcing feedback mechanisms, of amplification of deviations. On the other hand, positive feedback loops assume a new revision of the concept of causality as similar causes are not always will produce similar effects due to the possibility of amplification of the effects (Maruyama, 1963). The most interesting, for System Dynamics, of the different types of feedback loops is their combination in system modeling. So, we will know positive feedback loops amplify the effects, while positive feedback loops Negative feedback loops dampen or stabilize them.

In the Population example, representative of the evolution of the population of a species, the relationship between "population" and "number of births" generates a positive feedback loop, an amplification loop. The larger the population, the greater the number of births, and a new larger population, and a larger number of births, and so on. While the loop "population", "food", "birth rate" and "number of births" is a negative feedback loop as it contains a relationship of negative character between population and food. In a generic way, we can establish that when in a loop there is a An even number of negative influence relationships, said loop is amplifying in nature. It is a positive feedback loop. On the contrary, when the number negative influence relationships is odd, the loop is one of decreasing amplifications. It is a negative feedback loop. On the other hand, the character overall system, amplifier or stabilizer, tending to development or stability, will depend on the force (Maruyama, 1963) or domination of one over others (Aracil 1983) of the different feedback loops. Weick (1969) offers a series of more detailed guides for determining the global character of the system. Thus, the first of them would be that the character of the system.

It is determined by the loop that contains the greatest number of elements. If this loop is reinforcing, the system will be of the character of amplification of the deviations; if it is stabilizing, the system will contain the deviations. If this rule is not applicable, for example, for equal number of elements in loops of different types, we will apply a second rule. The character of the system will be given by the predominance of positive or negative loops, that is, by the predominance, in number, of a type of loops. A third rule to apply to the detriment of the preceding two consists of counting the total number of relations negatives between all the elements of the system, counting each negative relation each time it appears within a loop. If this total number is odd, the system will be of the stabilizing type, which counteracts the deviations, if the number is even, the system will be of the reinforcing type, which amplifies the deviations. Applying these guides to the population example, it is enough to take into account the first of the rules: the system is stabilizing in nature as to the greater number of elements which implies the negative loop versus the elements involved in the positive loop.

Equations, with their solutions, can also be worth many words.

The strength of a causal loop diagram is that it helps us understand how individual elements function together as a system. We can literally see the large-scale structure of the whole system. All the individual elements are clearly identified. How each element influences, or does not influence, each other element is also made clear. It allows us to trace through the effects of any one element on the whole system by following its links from that element to the next elements it is connected to, and from those elements to the further elements that they are connected to, and so on. In this way we can build up a quasi-dynamic picture of what would happen if the value of that first element were to change.

Often, sets of elements and their links form a cycle, that is, the trail of connected links return to an element from which they started. These cycles may be "positive feedback loops" also called "reinforcing loops" where the returning influence increases the value of the starting element.

> *"These cycles may be "negative feedback loops" also called "balancing loops" where the returning influence decreases the value of the starting element. These reinforcing and inhibiting feedback loops are given an essential role in understanding system properties as they determine how the flow of information or energy increases, or decreases, the values of elements or groups of elements. 2 A causal loop diagram serves many different purposes."*
>
> (Vandenbroeck et al. 2007).

> *"It is a heuristic tool that supports and promotes meaningful conversation among experts to develop new questions and hypotheses for data gathering and theory building. It is a knowledge management tool that organizes the available knowledge in an integrative way and illustrates how individual elements and groups of elements fit into large-scale structures of the system. It is a diagnostic tool that helps to identify potential gaps in current policy approaches. It is an operational tool to identify "leverage points" in a system as a starting point for policy interventions."*
>
> (Meadows 1999).

> *"Perhaps, most important, it is a process tool. The diagram can be created as an interactive team exercise that forces the team to think up and think through the essential elements of a system and how they interact with each other."*
>
> (Burns 2007).

> *"That is why it has been said that 90% of the value of the causal loop diagram actually resides in the process that created it, and only 10% in the finished diagram itself."*
>
> (Danny Burns, personal communication, 2014).

> *"This process allows the team to view and therefore frame the system from multiple perspectives. The ability to then merge these multiple perspectives into a coherent whole is a core aspect of "integrative complexity," which research has found leads to better decision making with complex problems."*
>
> (Redding 2016; Suedfeld 2010; Tetlock 2005).

A casual loop diagram also has limitations. Even though above we referred to the "value" of the elements, these elements have no quantitative, or even ordinal, values. It is also essentially a static representation of the interactions between the elements, while real systems display interesting and sometimes ever-changing, patterns in time. In principle, it might be possible to think through if this element changes, that element will change, which will change something else, and so on. However, in practice that is very hard to do when there are many elements. Moreover, the value of each element must depend on the values of all the other elements and their interactions, even though that value cannot be determined from the causal loop diagram. We will show below how these limitations can be addressed by developing a quantitative mathematical model from a causal loop diagram. (Navarro 2001).

Delays in Feedback Flows.

Along with the interest in feedback loops, either positive or negative, Systems Dynamics emphasizes the importance of delays or delays in the transmissions of information or material.... That is, when building a model using causal diagrams, it must also take into account that the influence relationship that unites two elements may require the course of a certain period of time.

For example, decisions to downgrade a product usually do not have an effect immediate in the number of sales of the same, but requires a certain period time to check these effects. Another example, between the realization of an order to the manufacturing plant and its receipt for sale is usually time will pass. In the latter case, Forrester was able to see how the delays produced in industrial distribution systems caused oscillations in orders (Forrester 1961), hence, from a managerial point of view, a greater control of delays leads to more control in the system. This idea that it is nuclear in distribution philosophies such as just in time (Navarro 2001).

About this workshop

At its simplest, the systems dynamics workshop seek to use rigorous computer based simulation models that create possible scenarios depending on managers decisions to show them possible unexpected fluctuations and work as a support prospective tool for the decision making process.

Building Systems Dynamics (SD) models with teams help to enhance team learning, foster consensus and to create commitment with a resulting decision.

SD can be helpful to elicit and integrate mental models into a more holistic view of the problem and to explore the dynamics of this holistic view.

It must be understood that the ultimate goal of the intervention is not to build a system dynamics model or a simulation.

Models are means to achieve other ends.

They put people in a position to learn about a messi problem, creating a shared social reality, a shared understanding of the problems and potential solutions to foster concensus within the team.

Vennix (1996)

The structure of a complex system is not a simple feedback loop where one system state dominates the behavior. The social complex system has a multiplicity of interacting feedback loops. Its internal rates of flow are controlled by non-linear relationships. The SCS is of high order, meaning that there are many system states. It usually contains positive feedback loops describing growth processes as well as negative, goal seeking loops.

Forrester (1969)
Causal models simulations and particularly SD models are a limited use when studying social phenomena. This is because causal models require modelled phenomena to be stable and determinate. If the constitution of social phenomena is inherently indeterminable and constantly subject to change and if causal models are unable to represent this constitutional change, then causal models can easily produce invalid results (Smagt 2006).

References

Ackoff RL. (1981). Creating the Corporate Future: Plan or be planned for. John Wiley: New York.

Ackoff RL (1999) Ackoff´s Best : His Classic Writtings on Management New York :Wiley.

Beer, S. (1959). Cybernetics and Management. English Universities Press, Oxford.

Beer, S (1974) Designin Freedom .Toronto :Canadian Broadcasting Corporation Publications.

Beer, S. Diagnosing the System for Organizations. John Wiley & Sons. 1985.

Beer, S. (1989) The Viable System Model: its provenance, development, methodology and pathology. in: The Viable Systems Model: Interpretations and Applications of Stafford Beer´s VSM (ed . R.Espejo and R.J Harnden) P 11–37 Chichester Wiley.

Beer, S. The Heart of Enterprise. John Wiley & Sons. 1979 International Campaign to Ban Landmines, www.icbl.org (last access Nov. 9 2021).

Bell, S. and Morse, S. (2013). How people use rich pictures to help them think and act. Systemic Practice and Action Research N 26 p 331–348

Braun, W. (2002). The System Archetypes. In: Braun, W., The Systems Modeling Workbook.

Checkland, PB and Poulter J (2006) learning for Action a Short definitive Account of Soft Systems Methodology and its use for practitioners, teachers and Students.

Checkland, PB and Poulter J (2010a) Soft Systems Methodology. In Systems Approaches to Managing Change : A Practical Guide mChicago, IL: University of Chicago Press.

Checkland, PB and Poulter J (2010b) Soft Systems Methodology. In Systems Approaches to Managing Change : A Practical Guide.

Checkland P (2003) Personal communication with Michael Jackson Jackson M 2019.

Espejo, R (2008) Observing organizations:the use of identity and structural archetypes. International Journal of Applied Systemic Studies N2–6

Espejo, R and Reyes A (2011) Organizational Systems: Managing Complexity with the Viable System Model Berlin Springer Verlag. estrategias/ diseños/ acciones. Ponencia presentada en la 1ra Jornada sobre Tecnologías Sociales, Programa Consejo de la Demanda de Actores Sociales-Ministerio de Ciencia.

Habegger Mancila J (2006) El poder de la Cartografía Social en las prácticas contrahegemónicas o La Cartografía Social como estrategia para diagnosticar nuestro territorio Sabina 14. abril 2006.

Huhn, M (2012) Cybernetics management paradigms:In Systemic Management for Intelligent Organizations.

Jackson M (2019) Critical Systems Thinking and the Management of Complexity. Responsible leadership for a complex world.

Ledington P & Donaldson J (1997) Soft OR and management practice : a study of the adoption and use of soft systems methodology. Journal of the Operational Research Socirtyu48: 229–240.

Meadows DH (2004) Limits of Growth : The 30-year update.

Midgley,G.(1997) Mixing methods:developing systemic interventions.In:Multimethodology (Mingers &Gill)p291–332 Chichester Wiley.

Navarro, J. (2001). "Las organizaciones como sistemas abiertos alejados del equilibrio". Tesis de doctorado no publicada. Universidad de Barcelona: España.

Oxford U (2021) Oxford University Press consulted 5 August 2021 https://oxfordre.com/education/view/https://doi.org/10.1093/acrefore/9780190264093.001.0001/acrefore-9780190264093-e-528.

Paulston R (1993) Social Cartography–What is Social Cartography?

Pérez Ríos, J. Diseño y diagnóstico de organizaciones viables. Un enfoque sistémico. Iberfora (2000–2008).

Ruegg-Sturm J (2005) The New St Galen Management Model: Basic Categories of an Approach to Integrated Management Basinggstoke: Palgrave Macmillan.

Schwaninger, M (2001). System Theory and Cybernetics: a solid basis for transdisciplinarity in management education and research. Kybernetes N 30 p 1209–1222.

Senge P. (1990). The Fifth Discipline. The Art and Practice of the Learning Organization. Doubleday/Currency: New York.

Ulrich W (1988a) Churchman process of unfolding its significance for policy analysis and evaluation Systems Practice.

Ulrich W (1988b) Churchman process of unfolding its significance for policy analysis and evaluation Systems Practice.

Ulrich W (2005) A brief introduction to Critical Systems Heuristics.

Ulrich, W & Reynolds (2010) Critical Systems heuristic in Systems Approaches to Managing change A practical guide.

Ulrich, W. (2012) Operational Research and critical Systems Thinking an integrated perspective , part 1:OR as applied systems thinking. Journal of Operational Research Society 63: 1228–1247, 1307–1322.

Study Cases 4

4.1 Private Companies

As we previously mentioned, the Untangling Complexity project has 5 phases, first the private companies where you can monitor and have direct control of most of the variables, then more complex systems such as public or mixed companies and then continue increasing the complexity by working with vulnerable communities in Colombia and then abroad to finally work with communities to build peace in Colombia.

The methodological proposal and the social transformation workshops were applied in private companies back in 2016. First in a German multinational H and then a Colombian private company M. The original idea was to implement critical systems thinking methods and methodologies in different complex social systems to analyze their effectiveness ans how easy it was for communities to use them. Once the most effective methods and methodologies for working with communities were identified, we were able to establish a process through which we controlledly reduced the complexity of existing within the complex social systems studied, this allowed decision makers who had the difficult task of making decisions that affect many people based on incomplete information and limited resources to make better decisions, in this case better decisions referring to decisions that will optimize the resources of the organizational systems of which they were a part to ensure the continuity and success of the projects they were in charge of.

C. A. Forero Navarro, *Untangling Complexity—Peace Building Engineering*,
Synthesis Lectures on Engineers, Technology, & Society 29,
https://doi.org/10.1007/978-3-031-37038-0_4

4.1.1 Case Study 1 H German Multinational Company

Case 1 Private Company 1 2016:

We began by applying the different methods and methodologies identified in private companies, complex social systems where we could make a detailed follow-up of the actors and variables and their interactions, and how, through the systematic application of social transformation workshops, we could reduce the existing complexity within the system in terms of obstructions, both discrepancies and conflicts, as defined by Russell Ackoff in his book Idealized Design. In this way, we can help decision makers to identify the key variables of the system, those that were under their control and that with adequate monitoring could allow ensuring the proper functioning of the systems they were in charge of, optimizing resources.

H is a multinational company headquartered in Düsseldorf, Germany. It is a global leader in the consumer goods industry, specializing in three business areas: Adhesive Technologies, Beauty Care, and Laundry & Home Care. Back in 2016 a first engineering intervention testing different methods of different critical systematic thinking methodologies was done. With a German private international company that operates worldwide with leading innovations, brands, and technologies in three business areas: Adhesive Technologies, Beauty Care, and Laundry & Home Care.

In the case of company H, they were in a process of organizational restructuring, entering the Latin American market, which generated various challenges and problems, generating problems and needs and a challenge for the decision makers of the company. The company was asked to apply the CFM Complexity Funel Methodology and the STW Social Transformation workshops to better understand the complex social system in which it was immersed and to be able to make better decisions. The private company System was identified as a non-viable system with many Organizational Pathologies. This meaning that not adaptable nor is it a flexible organization. This does not mean that it is not operational or working, but that it does not operate in an optimal way.

From our first experience of the systematic application of the social transformation workshops it was possible to see how they allowed gradually to reduce the complexity in the organization in terms of obstructions, both discrepancies and conflicts, thus helping decision makers to have a clearer image of the system in which they were working and could make more informed decisions, effectively accelerating the materialization of projects within the company. By identifying the existing discrepancies, spaces for dialogue were squared to share the different perspectives on the same point and some discrepancies were solved in the same way, conflicts were identified, and agreements were reached to continue casting despite these.

4.1.2 Case Study 2 M National Technology Company

Case # 2 Private Company # 2 2017

M is a Colombian online grocery delivery platform that provides convenient and efficient grocery shopping services to customers. It operates primarily in urban areas and aims to simplify the grocery shopping experience by offering a wide range of products that can be ordered through their website or mobile app. Customers can browse through various categories of products, including fresh produce, pantry items, household essentials, beverages, and more. They can add items to their virtual shopping cart, select delivery options, and proceed with the checkout process. M offers same-day or next-day delivery, allowing customers to receive their groceries conveniently at their doorstep. M differentiates itself by owning and operating its own fulfillment centers and employing its own delivery fleet. This vertical integration enables them to have more control over the entire supply chain and ensures efficient operations and reliable delivery services. The company also provides a subscription service called "M Prime" that offers additional benefits to subscribers, such as free shipping, exclusive discounts, and access to special promotions. M has gained popularity in Colombia for its convenience, reliability, and wide selection of products. It caters to the growing demand for online grocery shopping, particularly in urban areas where customers seek time-saving alternatives to traditional brick-and-mortar stores. In the case of company M, they were entering the Colombian market and wanted to maximize the number of new users and retain them, making sure not to lose them by offering them several advantages over their competitors. Back in 2016 a second engineering intervention in a Private binational technology company that operates under an online platform for home market delivery was done. It was born in 2017 in Colombia and currently operates 25 cities between Colombia and Mexico. This company allows users to market through its website or mobile application available for Android and IOS. In this, the user selects the products, schedules the delivery time in his/her preferred band and selects the desired payment method, this can be online or on delivery with a data phone or in cash.

According to its creators the company is the supermarket of savings not only in money, but also in time; since it allows its consumers to optimize the traditional model of supermarkets through a technological solution.

With 4 years of operation, this is a startup with more than 400 employees hired directly that has managed to change the traditional way of doing the market in more than 300 thousand homes in Colombia. In February 2018, it exceeded US $ 20 million in sales and achieved investments of US $ 14 million and again expanded its coverage to the municipalities of Madrid, Mosquera, Funza and San Luis (Colombia).

The supermarket has a stock of more than 5,000 products of the usual first-level brands in the traditional Colombian family basket. It works with a Low-Cost business model, manages its relationships directly with suppliers and manufacturers, centralizing the entire supply chain with its inventory in closed-door warehouses, from where the dispatch and subsequent transport of the products requested by the application is carried out or the web platform. This process works without any intermediation, which allows reduction in operating costs that translate into low prices for the final consumer.

The company defines its main strategy to revolutionize retail, having a presence in each of the consumer buying moments. With the current model, it covers the supply and replacement needs of consumers.

Work was done on the analysis of the user acquisition and retention system in this binational private company through the application of the Complexity Funnel Methodology CFM.

It was quickly identified that decision-making agents where resources, money and key decisions are concentrated for the operation of the organization.

There was an opportunity to provide information that allows better decisions in the company and improve the user acquisition and retention system.

Works well, but does not perform at optimal level, there is an opportunity for improvement.

Colombian entrepreneurship with great potential to offer a service needed in cities with accelerated rhythms of life and rising costs of living.

Generate meaningful recommendations that provide added value by improving M's operating and operating processes.

Structural Analysis: Systems Thinking Methodologies.

Simulation with experimental software: Dynamics of systems.

In our second intervention in private companies we observed similar results, through the systematic implementation of the different workshops we managed to improve the general understanding of the system, several obstructions were clearly identified, both discrepancies and conflicts that made it possible for decision makers to take measures by creating spaces for dialogue with the different actors to solve the discrepancies that arose from different perspectives and allow all the actors to be aligned. Conflicts were also mapped and agreements were established on these in order to give continuity to the project. We then again evidence that the engineering tools, mainly critical systematic thinking, project management and participatory research and action, were successful in intervening this type of complex social systems.

The decision makers of the company were then asked to explain to us how the workshops and the methodology helped them in their decision-making process.

They explain how the web of life and rich picture workshops help them establish a common model. Then how Viable System Methodology and organizational cybernetics help them understand the importance of ensuring the existence of level 1 recursion, consolidate the marketing area. They were able to identify different Organizational Pathologies in their organization like the lack of communication channels. They also identify the need to strengthen system 3. They use Interactive planning and Idealized design to create the new market research team—"Insights" team. Finally, they use the Systemic

Archetypes to understand the system behavior and leverage strategies that ensure healthy growth.

The effectiveness of the use of engineering tools to support the decision-making process in complex social systems was demonstrated.

In conclusion, after having identified all the actors from the private company applying organizational cybernetics, the viability of the system was demonstrated by testing compliance with the three fundamental principles. The principle of recursion, the law of the required variety and the presence of the five articulated systems. As the company has all the characteristics of a viable system, it is a system that can face the complexity of the environment. Once all the actors had been identified, it was possible to see how they are related to each other and to identify the variables that each of these actors' handle. Taking this perspective, a problem was identified. This is that the company has a good user acquisition and retention system, but being a young company, it still has areas for improvement and therefore it is proposed to make a recommendation to the key decision makers of the company to improve this system. Which is at the base of the growth of the organization.

Considering the five articulated systems found with the organizational cybernetics model, a one-by-one diagnosis was made, and its exact function was determined to see if it fulfills its role in the company. This analysis allowed the identification of three different types of organizational pathologies in the system. Structural, functional and communication channel pathologies. Considering these pathologies, it was possible to identify which are the most important to solve in order to improve the user acquisition and retention system in the company. The company is recommended to review and structure the one or operating system. Provide them with the necessary resources and ensure good communication channels so that the information that reaches them is already transformed so that they can more effectively execute what is planned for the organization. A strengthening of system three is necessary in the organization since it is in charge of distributing resources and guiding system two and one in its work. It is proposed to put the user at the center of all projects to solve the "headless chicken" pathology and train the entire company on its mission, vision and values to end "organizational schizophrenia".

After this diagnosis, a critical systems heuristic was applied that seeks to involve all affected actors and ensure the validity, legitimacy and continuity of the project. This was done by answering the questions proposed by the methodology to highlight the sources of control, knowledge, experience and legitimacy. By involving those affected in this way, it is intended that this project be a process of co-creating a solution together with those affected and generating empowerment in the company operating units. Once this boundary was defined, it was possible to proceed with the rest of the structural analysis.

To improve the model of acquisition and retention of users in the company, it was decided to use the interactive planning model and idealized design, which allow us to adopt another dynamic in the face of the problem. For this, the creation of a team in the organization focused on the organic growth of the company was used. With this team, a tool was designed that made it easier to reach the idealized model. This tool consisted of three focus groups to understand the needs of the users. This also gave a solution to one of the aforementioned pathologies. Once this model was determined, it was possible to establish goals, objectives and goals to arrive at the idealized model and the difference or discrepancies with respect to the current model of user acquisition and retention were looked at. From there, two projects were born to arrive at the idealized model.

The first project is the consolidation of a team that focuses on the study and understanding of users to predict their purchase intention and be able to personalize the experience with the company. The second is the consolidation of a team that focuses on the study and experimentation with the user conversion funnel in the company to improve all conversion rates between each stage. This would allow the organization to acquire and retain more users. The final recommendation is to create both teams.

To carry out the application of the experimental model, it was decided to make an example of a case study of the conversion funnel optimization team, which should generate test scenarios to see how it behaves when certain parameters change. In this case, he simplified the Funnel model and experimented with the variation of the pressure element on the equipment to obtain results. Three scenarios were tested for this exercise. A base scenario, a pessimistic one and an optimistic one. With the results we were able to realize that maintaining constant pressure on the teams regardless of the level of growth of the organization, is better than varying this pressure. This is aligned with one of the pathologies mentioned and is that of the lack of a recursion level that allows transforming the requirements or this pressure on the operational units so that they can better develop their work. These results reiterate the recommendation to create the proposed teams because they show significant improvements in the user acquisition and retention model.

Finally, different systemic archetypes in the organization were studied to see which of these anti-intuitive behaviors were occurring in the company and could be avoided. In this exercise, three archetypes were found in the company. The first is escalation, the second is failing solutions, and the third is growth limits. The three archetypes have to do with the role of the pricing strategy used to acquire users and the organization's responsiveness to peak demand. The recommendation is to redefine your pricing strategy by identifying another competitive measure, evaluate the possible secondary effects of the proposed solutions to ensure that in the long term they will not affect the organization, and set growth goals in certain activities so as not to incur cost overruns. Logistics when there are peaks in demand.

Taking into account all these recommendations, The Company is expected to improve its user acquisition and retention system, giving it the ability to have greater control over its growth as a company. This organization is a Colombian enterprise that has all the tools to become an international business due to the need for the service it offers. The rhythms of life around the world accelerate daily and make activities as basic as getting groceries a very time-consuming and effort-consuming task. People's time is a very valuable resource and by taking this into account in its value offer, The Company is positioned as a very promising company in the technology sector.

The second application of the multi-methodological proposal of the Complexity Funnel Methodology CFM and the Social Transformation Workshops STW showed us that in complex social systems of private companies where it is easy to follow up on actions, a reduction in complexity was achieved in terms of discrepancies and conflicts within the studied system, important and in a short time, helping us to determine the effectiveness of the use of certain methods and methodologies and making our proposal more robust.

4.2 Public Companies

4.2.1 Case Study 3 Colombian N Bank

6. Case 3 Public Company #1 Colombian Central Bank

We Start the second phase of the Untangling the complexity project intervening-Colombia (BRC) is A bank of Colombia, in charge of issuing, managing and controlling the monetary movements of the country, as well as issuing the legal tender, the pesos. The Colombian N Bank, also known as Banco de la República in Spanish, is the central bank of Colombia. It is the country's monetary authority responsible for maintaining price stability, regulating the financial system, and promoting the sound functioning of the Colombian economy. In the case of the Banco Nacional de Colombia company, we sought to understand when the bank allowed other banks to increase their usury rate when the economy was bad. But we wondered what happens when the economy improves as the banks seem to have no incentive to lower rates again at least for a while.

The project with the N bank in 2017–2018 was called the Analysis of the banking system in Colombia in search of new strategies that promote a change in it.
The problem arises due to the lack of coordination and control of the Colombian banking system, which generated a resistance in the agents to follow the policies of the state. Consequently, and with the passage of time I end up in a disorientation in the system. This means that the system does not fulfill its real purpose, which means that instead of prioritizing the preservation of value for citizens, the generation of value for the bank is sought. Thus, achieving an unreliable central bank without decision-making power.

"*The reduction in the intervention rate of the Banco de la República has not been reflected in consumer loans as would be expected to have an impact on the activation of the national economy*"

(The Herald 2018).

The true justification, as we can see in the phrase written by the herald, is the problem that is generated when the state and the bank seek economic activation. Not having policies that I force or coordinate the agents to adopt and work for the purpose of the banking system is affecting citizens, especially because it is taking away the value of their money as they lose sight of inflation and this delay in rate adjustments. they cause a slow return to a more desired economy.

Different intervention strategies or projects were proposed to align the behavior of the agents with the purpose of the system, based on different methodologies that make the system more effective.

Find different projects that influence the banking system, thus maintaining inflation at less than 7%.

Understand the behavior of the interest rate and the agents to change the intervention rate of the bank of the republic.

Identify which interactions and agents are important to direct strategies or projects and generate a greater impact on the alignment of objectives between entities.

"The banking agents prioritize their personal interest over that of the system, leading to them not responding to the strategies chosen by the government and the bank of the republic to control or coordinate them."

We observe how, unlike the interventions in private companies, unfortunately in public companies the organizational entropy decreases much more slowly, this is evidenced in the 3d model of the intervention.

Integrating the results of the three methodologies plus the analysis of archetypes in the system, we conclude that projects need to be carried out in the political, banking and educational areas to generate a change in the banking system.

In the political area, it is recommended to generate the following political reform project, which consists of creating a control entity independent of the state to monitor assets, the creation of a citizen control entity and a tax reform.

In the economic and financial area, it is suggested to integrate professionals from different areas in the different organizations that promote a different thinking than the economic one for decision-making, create communication channels between the academy and the banks so that the investigations they carry out have an effect on the decision of the banks and gradually eliminate transaction costs, especially interbank ones that promote real competition.

Finally, in the educational area, both the government and the banks have to generate publicity and new financial products, create financial communication campaigns that have up-to-date and easily accessible information and create a free financial education project to try that people can take better decisions whether as borrowers or as depositors.

In the third application of the methodological proposal of a complexity program and the social transformation workshops, we realized that in public companies the reduction of complexity in terms of discrepancies and conflicts within the systems is not so easy. Unfortunately, the bureaucracy that exists within these makes the process slow, but in a much more delayed manner.

4.2.2 Case Study 4 Cartagena Mare

Mayor's office of the city of Cartagena
Cartagena known since the colonial era as Cartagena de India's is a city and major port on the northern coast of Colombia in the Caribbean Coast Region. Founded in 1533, the city's strategic location between the Magdalena and Sinú Rivers gave it easy access to the interior of New Granada and made it a main port for trade between Spain and its overseas empire, establishing its importance by the early 1540s. The lionfish (Pterois volitans and Pterois miles) is an invasive species that has become a significant problem in the Caribbean coast of Cartagena, Colombia, and other parts of the Caribbean region. Originally native to the Indo-Pacific, the lionfish has been introduced to the Caribbean through aquarium releases or accidental releases from aquaculture facilities. The lionfish invasion is considered a problem because the species has no natural predators in the Caribbean, and its population can rapidly expand, negatively impacting the native marine ecosystems. In the case of the lionfish in Cartagena, the mayor requested our services to make a simulator based on incomplete information on the lionfish problem to be used for the construction of a public policy. Initially, a proposal was presented where local fishermen were paid for catching lionfish and local restaurants for offering lionfish dishes. At first the initiative seemed to work very well as the number of lionfish was decreasing thanks to the activity of local fishermen who could then sell them to restaurants. However, as the population of the invasive species decreased, the fishermen did not receive payments for the restaurants, the restaurants did not have raw materials to work with, and the economic model that had been created around this invasive species collapsed because the fishermen did not receive pay they found other jobs in the tourism and transportation sector mainly and restaurants stopped offering their dishes. As there were no fishermen, the invasive species began to increase again, putting our local fauna at risk.

Modern Cartagena is the capital of the Bolívar Department, and had a population of 1,028,736, according to the 2018 census, making it the second-largest city in the region, after Barranquilla, and the fifth-largest city in Colombia. The urban area of Cartagena is the sixth-largest urban area in the country, after Bucaramanga. Economic activities include the maritime and petrochemical industries, as well as tourism.

The Cartagena Mayor's Office and the Corporation Autónoma del Canal del Dique (Cardique) launched in 2017 a campaign in the city's beach areas to alert bathers, tourist servers and fishermen about the presence of the dangerous lionfish in the waters of the Caribbean Sea. The capture of several of these animals in Lomita Arena and a bather's complaint of having been attacked by a lionfish, forced the authorities to take preventive measures, considering the high danger that the bite of this natural predator has. The Mayor's Office and Cardique started the prevention campaign on beaches, hotels, docks, and populated centers that carry out fishing activity, to warn bathers about the lionfish, an invasive species that threatens ecosystems due to its voracity and the health of humans due to poison that they have on their dorsal fins. According to The Mayor's Office and

Cardique's director, Olaf Puello, the call is not for people to leave the beaches, but to be careful if they observe the animal near them. "This fish is doubly dangerous because, first of all, it is an active predator of crabs, mollusks, shrimp and small fish, affecting reef processes and the biodiversity of marine ecosystems. In addition to that, it affects the economies of coastal sectors that live off tourism and there are the effects it causes on humans as a result of the poison associated with the spines of its showy fins. "If a person suffers a puncture by this fish, paralysis of the affected limb and adverse effects on the cardiovascular system, among others, can occur," said Puello Castillo. According to The Mayor's Office and Cardique's director, if a person finds a lionfish they should refrain from handling it, take note of the location and notify the environmental authorities. In case of being attacked, immediately apply hot water that is supported and visit the nearest health center to receive medical attention. Lionfish venom remains active for up to 30 min after the animal's death. It has cardiovascular, neuromuscular, and cytolytic effects, ranging from swelling to extreme pain. Several bathers claimed to see several of these animals prowling the beaches of Bocagrande. Mayor's Office and Cardique's campaign extended to El Laguito, Castillogrande, La Boquilla, Punta Canoa, Marbella and Crespo, among other beaches.

The Cartagena mayor's office decided to ask for help to make some simulation models that would allow to support the decision support system regarding the lionfish problem.

Not all engineering interventions are successful interventions. It is also important to document failed interventions. This is the case of the intervention carried out with the Cartagena mayor's office in the Leon fish project. It is a project that seemed to be successful in terms of reducing the complexity of the system, but the fact that the mayor's office could not ensure the continuity of the project over time created the perfect conditions for an even more complex system with even worse difficulties. This is an example of an archetype of quick fixes failing. Technological tools were used to help decision-making and try to solve a problem, however, the partial understanding of the problem made the solution given only a symptomatic solution but not a fundamental solution to the problem.

In the fourth case of application of the multi-methodological proposal of the complexity tunnel and the social transformation workshops, we had a case of failure due to the client's need to use robust engineering tools such as simulators without really understanding the variable actors and their interrelationships. This is very important because it generates relevant knowledge about the things not to do when doing an engineering intervention and Seeking to use very robust tools without having a clear understanding of the emerging properties, interrelationships, and elements of the systems. Unfortunately, in this case, the use of innovative and robust tools did not produce the expected result but inadvertently generated the perfect conditions so that the original problem would be even more difficult to solve. This generates a very important lesson that is one of the key conclusions of this work and that is that the identification of the engineering tools that must be used in the different types of systems is key, since the good identification of the appropriate tool is what will allow us to give an effective, efficient and effective solution to the needs and problems that arise in complex social systems and in opposition the misuse of tools in these systems inadvertently and intuitively can generate the perfect conditions. s for problems to escalate and become even more difficult to solve.

4.3 Communities

4.3.1 Case Study 5 Venezuelan Migrants

After working with private companies and public and mixed companies we tested the implementation of social transformation workshops with communities in Colombia (Fig. 4.1). The influx of Venezuelan migrants into Colombia has presented various challenges and complexities for both the migrants and the Colombian government. In this case, we work with various groups of the migrant population who were given seed capital and support to start some type of productive enterprise. We observed curiously that the ventures that were related to Venezuelan gastronomy were very successful and easy to replicate, offering business opportunities for these population groups.

The arrival of hundreds of thousands of Venezuelans to Colombian territory in the last decades has ceased to be a novelty to become a social and economic phenomenon in Colombia. The complex political situation that Venezuela is going through is the main cause of migration, which at the beginning was mostly made up of highly qualified people who have been successfully inserting themselves in various productive sectors. Although

Fig. 4.1 Complexity Funnel Methodology and Social Transformation Workshops Public Company 2 Int 2016 Industrial Engineering Camilo Navarro University of los Andes Bogotá DC Colombia

the latter was positive then it was a middle class that had difficulty and at the moment we are already receiving people who come to the country with nothing. Migrants are resilient people and many seek opportunities to get ahead in informal sectors of the economy.

A very interesting result is the decrease in the time of reduction of the organizational entropy. Although the organizational entropy is reduced as in the cases of private companies and public companies. It takes much longer when we worked with vulnerable communities.

This is an interesting result for our thesis because it shows that although the engineering intervention is effective, it takes much longer to take effect. This creates other possible questions such as what are the reasons for this time delay and what can we do to solve it.

In our fifth case of application of the multi-methodological proposal of the Complexity Funnel Methodology and the Social Transformation Workshops, this time with vulnerable communities in Colombia, more precisely with migrant communities in the city of Bogotá, we realized that, as in the public company, change is possible but it takes much longer due to the difficulty that exists at the time of implementation and monitoring. Change occurs, but it occurs much slower than what we had observed in private companies where we could permanently monitor and accompany the actions and the decision-making process.

4.3.2 Case Study 6 Syrian Refugees

Syrian migrants in Canada face various challenges and experiences as they settle into their new lives. While it is essential to approach the topic with nuance and recognize that experiences can differ among individuals, Back in 2015 Canada's response to the humanitarian crisis in Syria was to welcome more than 25,000 Syrian refugees by the end of February 2016. The Government of Canada or the province of Quebec provided support. It lasted for 12 months or until they were self-sufficient, whichever came first. Support covered items like:

- clothing
- basic household needs
- staples
- linen
- furniture
- winter coats
- boots
- food
- shelter.

During their first weeks Canada, Syrians, like all resettled refugees, needed help with immediate and essential services like:

- getting a social insurance card
- opening a bank account
- learning how to use public transportation.

A program called the Resettlement Assistance Program helped refugees with these tasks.

Privately sponsored Syrian refugees went to live in their sponsors' communities. They received similar support from their sponsor.

Syrian refugees also received medical coverage under the Interim Federal Health Program.

Like all permanent residents, Syrian refugees also had access regular settlement services, like:

- language training
- help finding a job
- community support.

Once they were organized with the basics, the Canadian government tried to support different business initiatives for refugees in their new country with seed capital. Curiously, very similar to the case of Venezuela and Mexico. The initiatives and projects that revolve around food, Mexican tacos, Venezuelan arepas and typical Syrian food were more successful. Offering real possibilities of employment, growth and development to these vulnerable communities. In the case of Syrian refugees in Canada, very similar to the case of Venezuelan migrants, it was observed that several projects related to Syrian gastronomy were very attractive and successful in Canada, since they represented a new gastronomic offer. We also observed that they were projects with great potential and growth prospects that generated employability and were easily replicable. Offering real opportunities for insertion into Canadian society.

In the fifth case of application of the multi-methodological proposal of the tunnel of complexity and the social transformation workshops, we once again observe that when applying the methods and methodologies of critical systemic thinking in vulnerable communities, the results come but take a long time compared to what we had observed in scenarios where we had more control and where we had the opportunity to monitor more carefully and closely the actions and the decision-making process of the bureaucracy. The lack of resources and the impossibility of being able to carry out permanent monitoring lengthen the time to reduce complexity earlier in terms of obstructions, discrepancies and conflicts, making the process of decision-making and solving the needs and problems existing within these complex social systems slower.

4.3.3 Advancing Students Forward Tijuana Mexico

In San Diego United States and Tijuana Mexico, we had the opportunity to work with the binational non-profit ASF. Advancing Students Forward-A su Futuro, the mission of Advancing Students Forward is to provide underserved students access to education and support for academic success so that they can create a promising future for themselves and for the community in which they live. ASF is a binational nonprofit, supporting students in Tijuana, Mexico and promoting cross cultural and educational exchanges between students on both sides of the border.

In Mexico 50% of students do not have the opportunity to continue their education beyond 6th grade. After 6th grade students are required to pay tuition and have the added costs of school uniforms, books and supplies. In addition, the number of schools is limited and many students are forced to travel outside of their community to find a school that has space for them. Transportation costs increase the financial burden and parents are hesitant to send their children alone on public transportation a significant distance away from home.

Advancing Students Forward (ASF) was founded in 1984 by Monica Santos who began volunteering at an orphanage on the outskirts of Tijuana, Mexico while a college student at Loyola Marymount University.

According to the Mexican prime Minister only 62% of students reach secondary school and at secondary school, about half of the students drop out. The educational goals of our students are impeded because of poverty, crime, and lack of school facilities. The number of students who are longing for an opportunity to go to school is staggering.

To address the disparity in access to education, Monica established the program informally by mentoring one student.

Our first student was given a manual typewriter and completed middle school. The next student became a preschool teacher. As time went on, word of the program spread, donations increased, students from neighboring communities were invited to participate and our organization grew to what it is today. Each year ASF serves around 70 students studying in middle school through the university level. Most of ASF students continue in the program for an average of 9 years. ASF also has an active group of alumni who mentor the current participants.

In the early years the primary goal was for students to complete secondary school (9th grade). Over the years support has grown and students have been given the opportunity to extend their studies. As long as a student continues to fulfill the requirements of the program they may continue participating in ASF.

In 2013, Students brainstormed and created the organization's logo and the Spanish name for ASF, "A Su Futuro" which means "*On to Your Future*". This process enabled the students to "*take ownership*" and identify in a personal manner with the organization.

When working with communities internationally, we realize that the results are similar to those found when working with communities in Colombia. It is observed again that the systematic application of the social transformation workshops reduces the complexity presented in the system thanks to the creation of a common model that allows us to identify the variables of interest and the existing obstructions. It also allowed us to generate spaces for dialogue to discuss the discrepancies found and see what actions could be taken to deal with them.

In the case of Advancing Students Forward we used the CFM complexity funnel methodology and the STW social transformation workshops to understand the complex social system in which the binational foundation was immersed, its members the children of the El Tecolote neighborhood in Tijuana, their parents family and other actors in the system, both the public-private sector and educational institutions, etc. This allowed us to have a holistic idea of the system of obstructions, both conflicts and discrepancies that existed within it and allowed decision makers that existed within the system have a holistic vision of the needs and problems of the different actors and make more informed decisions.

In the sixth case of application of the multi-methodological proposal of the complexity Funnel Methodology CFM and the Social Transformation Workshops, STW we observed that the creation of safe spaces for dialogue and the implementation of these methods and methodologies systematically reduced the existing complexity within these systems in terms of obstructions, discrepancies and conflicts, helping in the decision-making processes within these complex social systems. Vulnerable communities in Colombia and vulnerable communities in Canada, we therefore recommend that when working with a vulnerable community you have the accompaniment of the academic part, since this will help accelerate the expected results by facilitating the decision-making work of the heads of these systems.

4.4 Peace Building Communities in Colombia

As the final stage of our project and after having had the opportunity to implement and refine the different workshops of social transformation, we approached different peacebuilding communities among which we have civilian victims of the conflict: The mothers of the first false positives in Soacha through the MAFAPO Corporation. Military victims and their families through the Matamoros Corporation. Demobilized from the AUC. People in the process of reincorporation of the FARC-EP. In the case of the peacebuilding communities and after testing the different Social transformation workshops in private companies, mixed public companies in communities at a national and international level and applying the different lessons learned, the CFM complexity funnel methodology was applied and the STW social transformation workshops with peacebuilding communities seeking to support productive community projects so that they are successful and offer real alternatives for reintegration into society for all the people who had been victims of the Colombian armed conflict.

4.5 Case Study 7 Cerveza la Roja

Case #12 Cerveza la Roja Peace Building Community 6

The signing of the Peace process by the Colombian government with the Revolutionary Armed Forces of Colombia EP (FARC-EP) in 2016 allowed some of the former combatants to lay down their arms and return to civilian life. A group of ex-combatants was assigned to the Antonio Nariño territorial space located in Icononzo, Tolima and a beer project arose there artisan that is known today as Cerveceria La Roja. In 2020 its production grew to the point who were able to move to Bogot'a and found their brewery plant (currently in operation in the Siete de Agosto neighborhood). In the case of the Cerveceria de Cerveza la Roja Cerveza Artesanal. The Funel de Complejidad methodology and STW social transformation workshops were applied with a cooperative of people in the process of reincorporation in the Antonio Nariño territorial space in Icononzo Tolima. From the application of these workshops, the idea of growing the business and having a brewery in the city of Bogotá to obtain an INVIMA certificate to be able to export Red beer abroad.

The process of disarmament, demobilization and reincorporation of former combatants is complex and requires compliance with guarantees by the National Government. After signing the Peace process, 13,000 people laid down their arms to begin a gradual process towards civil life and this underscores the importance of support programs and guarantees so that these people can develop personally and professionally.

So, this project aims to work with one of the communities of ex-combatants so that the reincorporation process is effective. Industrial engineering covers a lot of knowledge and aims to intervene organizations and systems to solve problems. The work has two pillars that work in parallel: the first consists in using the methodologies of Social Transformation Workshops and Complexity Funnel (Navarro and Dalrymple 2019) to form a multidisciplinary dialogue between agents and actors of the system so that projects that correspond to the community needs. The second pillar corresponds to the materialization of the projector prioritized from the co-creation between the academy and the community.

For more than 60 years, the Colombian territory has been a focus of violence. Two conflicts strongly marked history in the twentieth century. In 1848, the assassination of the presidential candidate Jorge Eliecer Gait'an detonated the Bogotazo and began the period of Colombian history that we know today as La Violencia. This era was characterized by having confrontations between political parties (Liberals versus Conservatives) since when a party was in the power completely excluded the other and this led to confrontations and attempts coup. Between 1948 and 1953 it is estimated that there were about 200,000 dead. With the goal of ending bipartisan violence, between Liberals and Conservatives agreed that they would alternate power and divide the positions public in equal parts. This agreement was called the National Front and covered the period between 1958 and 1974.

Although the National Front contributed to reducing bipartisan violence, the persistence of structural factors such as income inequality, institutional weakness and disputes of land promoted the emergence of the conflict present in the country. On the other hand, the Front Nacional did not generate participatory spaces for parties of different tendencies to those of the two traditional parties and this concluded in the creation and subsequent consolidation of groups guerrillas and insurgents who settled in different regions.

The FARC-EP, one of the most important revolutionary groups in Colombian history, began with its founding in 1964 when the Liberals in Marquetalia had armed themselves to confront the Conservatives and did not surrender their weapons even when General Rojas Pinilla had granted a general amnesty. After being attacked, the guerrilla was founded mobile and remained active and growing until 2016. Between those periods, various dialogues were held seeking the disarmament of the group outside the law, but it was until August 26, 2016 that the Peace process between the Colombian government and the FARC-EP was signed or signed.

It is important to mention that the 2016 Peace Agreement has not been the only attempt in Colombia for the construction of Peace. Since the 1950s, an attempt has been made to open a dialogue with groups outside the law. However, many of these processes failed and it was until this century that took a real breakthrough. In 2011, the Victims and Land Restitution Law was created as a search to create normative parameters to build a real and lasting Peace. The focus of all agreements are truth, reconciliation, justice and the search for a reconstruction in the social fabric of the country. In 2012, the Legal Framework for the Peace that proposes a mechanism to reduce sentences and facilitates negotiations. With this, begins a rigorous process that concludes with the signing of the 2016 agreements. Relevance this agreement is enormous and allows us to understand that there are new dynamics and processes that must be studied from all areas. From this, the Peace Building Engineering, which is designed to legitimize and create continuity with the Peace processes.

During these five years, after the agreement was signed, the National Government has promoted the protection of the FARC-EP community while they conclude their reincorporation process based on new projects, companies, and undertakings. However, many of the former combatants did not have a place to travel to after the war, so they formed communities in the Territorial Spaces for Training and Reincorporation (ETCR). There, training and early reintegration activities are carried out with which it is intended facilitate the initial phases of adaptation of the members of the FARC-EP to civil life. Equal form, it is intended that these actions can contribute positively to the communities surrounding. Below is a map with the location of the 24 ETCRs.

Initially, it was sought to mobilize ex-combatants to remote rural areas in a way that the adaptation process began little by little. Later, this population would leave the areas to return to the cities or towns to which they belonged, but many of them do not they had a place to return, so they created new communities in these areas granted by the government thus creating one of the first discrepancies. Despite having been planned, the agreements failed to foresee what would happen, and it is for this reason that they have arisen more problems, and therefore more conflicts and discrepancies.

Although the National Government has ensured the protection of the communities of ex-combatants, a fundamental and forgotten axis has been the continuity of the processes. The zones territorial areas have had problems due to lack of resources since the Peace Accords included economic aid that allowed the creation of jobs and entrepreneurship, but the help did not always arrive. However, the communities are very hard-working and have achieved create these growth spaces themselves: they have created textile workshops, livestock groups, breweries, tourist hostels, clothing manufacturing and weaving areas, among others.

In addition, the government's protection guarantees have not been sufficient and it is seen that there are a huge security breach for leaders, who are often being assassinated by illegal armed groups (Ball et al., 2018). For this reason, not only is the attention of the government (which proved to be insufficient) if not that the academy works to support the Peace process through its multiple tools to empower the communities to their growth and avoid further stumbling in the process. For this reason, it is very important that from the different disciplines apply knowledge in projects and methodologies that allow building and advancing on an issue so important for development and violence reduction.

This is essential now that we are in a post-conflict period.

La Roja Brewery is a craft brewery created by a group of ex-combatants of the FARC-EP. The information presented in this document and in the workshops is based on the compilation of dialogues with the head of the brewery: Carlos and Adriano.

The history of La Roja stems from the breach of the Peace Accords by them national government of Iv'an Duque to the extent that the agreement stipulated that the government promised to finance the productive projects of the ex-combatants so that after two years of laying down their arms they could have a productive project that allow work to have a decent life.

This has not happened after 5 years and to that extent in Antonio Nariño's ETCR.

In Icononzo, Tolima they began to look for alternatives for financing and production starting from their effort instead of waiting for government resources.

After contacting a friend from the community, Wally, they decided to do the course to learn how to make craft beer. Once the community realized that the process making beer was not an impossible thing to do, they began to learn and appropriate the draft. With that, they began to produce and market the beer with close acquaintances. and when they saw that people began to buy it, they decided to continue with the process by fine-tuning the recipe and production.

With the money reinvested from beer sales they have been able to increase production, by Initially they produced 25 L in a batch (20 days), after six months 250 L in a batch and they have gone growing until today they have the plant and have 4,800 L of production capacity. To Figures that show the original floor plan in a house in Icononzo are presented below.

Once they decided to invest their resources in beer, they founded the brand that is known today as La Roja and registered the trademark with the Colombian Chamber of

Commerce and the DIAN. This process was particularly difficult and time consuming as the community did not have counseling legal to accompany them with these procedures.

Once they got the trademark registration, they rented a plant close to the ETCR to produce and market the beer. However, the dream was to produce it themselves. With that idea, they had a visit from INVIMA at the ETCR and after a meeting the conclusion was that it was not possible to have the plant there due to the property rights of the land (not being of they couldn't build the plant unless they changed the POT). This is very important since the INVIMA permit is the one that allows them to market.

After this, they decided to make the financial effort to move the plant to Bogota and with this access to the INVIMA registry. With an arduous search, they ended up installing themselves in the Siete de Agosto neighborhood and seeking financing from the cooperatives, they bought the machinery (the Financing was difficult since, as they did not have a credit history, they were not selected for credit for part of the banks). With the machinery and the space, they still needed a significant investment: rebuild the space to transform it into a brewery plant.

Below are some images of the plant as soon as they arrived and after adjustments:

Currently, they have a production plant that only produces 50% since without the INVIMA permission cannot make wholesales or sales to establishments. East year they were able to start with new ideas that allow to increase the sales projection. Through the international organization UNDP (United Nations Program Development) founded a Pub at the front of the plant as presented in the following figure.

The community work has been very expensive, and they currently have a process of sophisticated production that rivals other craft breweries. They currently carry out the production of two types of beer: La Roja classic (red) and La Roja Rubia (golden). What's more, they lease two of their machines to other breweries that require them.

La Roja is just one example of the difficulties faced by people in the process of reintegration. For this reason, it is relevant and important to support the communities that are in reincorporation so that a lasting Peace can be built in Colombia.

Based on the new reincorporation problems that have arisen since the signing of the Peace process, new disciplines have made their way to try to understand and solve the conflicts left by the Peace Agreement. Engineering has carried out projects that support the post-conflict. An example of this is a job that optimizes the organization of the system Colombian criminal law, articulating it with transactional justice through a model that simulates the way the system is organized and the way some actors relate to others. With this information, the simulator implements different types of decisions: what would happen?

If more prosecutors were assigned, what would change if the judges were placed in another stage of the process or what would happen if the inflow of criminal news increases or decreases. Thus, it is examined systemically the consequences of possible interventions.

Another example of social projects from Industrial Engineering is Engineers without Borders.

The project seeks, through education, teaching and research, to motivate students to develop proposals with social impact. 1,000 students (approx. 700 Uniandinos), 25 teachers (12 from Uniandinos) and more than 2,000 high school graduates have been part of this organization (Cajiao et al., 2019). In addition to the above, several projects developed since the academy to support the Peace process. For more information click here.

On the other hand, methodologies have been developed that allow the application of Industrial Engineering for Peace Building. In particular, an example of this is a one organizational heuristic that

(1) makes an analysis of the system to understand it in its entirety,
(2) use appropriate tools to solve the identified problems and
(3) implement a solution capable of lasting over time (Navarro and Dalrymple 2019). The metric is called Function of Complexity and it is from Social Transformation Workshops (TTS) that it is achieved evaluate and decrease the complexity of a system (also called organizational entropy). Taking into account the above, the importance of contributing to social processes is highlighted from the academy and in particular to the Peace process supporting the community of ex-combatants of La Roja.

From the results of the Social Transformation Workshops, it was possible to understand the complex system Cerveceria Artesanal La Roja. With this, the entropy was calculated organizational at each moment of time and used the metric of the Complexity Funnel to.

See how clutter was reduced in the system. This with the aim of providing greater information to decision makers.

The system is currently unitary, so it is already in the last stage of the Complexity Fynnel Methodology. With this, the materialization of projects occurred from the co-creation between the members of the community and thesis students. However, the system still has a complexity given by conflicts and discrepancies that have not been resolved (such as export permits).

This productive project strongly aims at the reintegration of several ex-combatants and it becomes relevant to continue supporting them to continue building Peace in Colombia. Expected that the beers of La Roja reach the whole world and A toast for Peace!

Due to the Covid 19 pandemic, the sale of products from cooperatives of people in the process of reincorporation were affected. To support the sales of La Roja beer, Virtual Tastings were organized in which people bought their beers at home and connected virtually with brewmaster Wally Broderick.

With the money collected from the sale of the red beer and the virtual tastings, donations were given to 3 ETCR so that associations such as ASOMUFAN, the Farianas Women's Association, could coordinate the creation of the construction of kindergartens and schools.

4.5.1 Case Study 8 Tej Paz Avanza

Case #11 Tejiendo Paz Peace Building Community 5

Articulation of Tejiendo Paz cooperative proposals

In Colombia, after more than fifty years of armed conflict with the FARC (Revolutionary Armed Forces of Colombia), the largest guerrilla in the country, the national government decided to establish an agreement with the purpose of negotiating peace. According to the primer of the final agreement of the Foreign Ministry (2016), by means of this agreement the guerrillas were expected to hand over all their weapons to the United Nations, not to commit crimes, to break their ties with drug trafficking and to cease attacks on the Public Force and the civilian population. In addition, in order to promote the reintegration of the FARC, a comprehensive agrarian development plan with access to land and services and a strategy for the sustainable substitution of illicit crops was proposed within the agreement. On the other hand, an investment plan for the countryside was agreed with programs for access to land, goods, productive services and infrastructure to give farmers real opportunities for development and quality of life (Fig. 4.2). In the case of the Tejpaz cooperative, it began with a sewing machine in the Antonio Nariño territorial space in Icononzo, Cundinamarca, where Don Gonzalo began to teach other members of the community how to weave and they began to produce their own textile products. They have had the opportunity to make alliances with university institutions and government entities that have supported the development of Brands such as "Manifesta" and "Avanza" and that step by step are offering real work alternatives for the members of these cooperatives and gradually facilitating and despite many difficulties the hard process of reintegration into society.

After the agreement, the initiative was born to support the reincorporation process for the peace signatories. This initiative consists of promoting the hiring of ex-combatants, as well as their undertakings, with the purpose of closing the economic gap that exists and enhancing the reintegration process. As a result of this, the Transitory Pathways for Normalization Zones (ZVTN) were created, today known as Territorial Spaces for Training and Reincorporation (ETCR). These are areas that were disputed by different groups during the armed conflict, and now they want to reuse as spaces that soften the adaptation to civil life for ex-combatants. Within the paths known as (ZVTN) is Icononzo, a rural area, located in the department of Tolima. The Tejiendo Paz Cooperative is located in this specific area, founded and operated by peace signatories in Colombia.

Having said this, the following document has the purpose of using engineering tools, specifically the Complexity Funnel Methodology and the Social Transformation Workshops created by Professor Camilo Navarro at the Tejiendo Paz Cooperative. With the use of these tools it is expected to reduce the complexity of the social system in order to make better decisions within it and thus support the reincorporation process of this group. The relevance of this project lies in the opportunity to promote the development of the

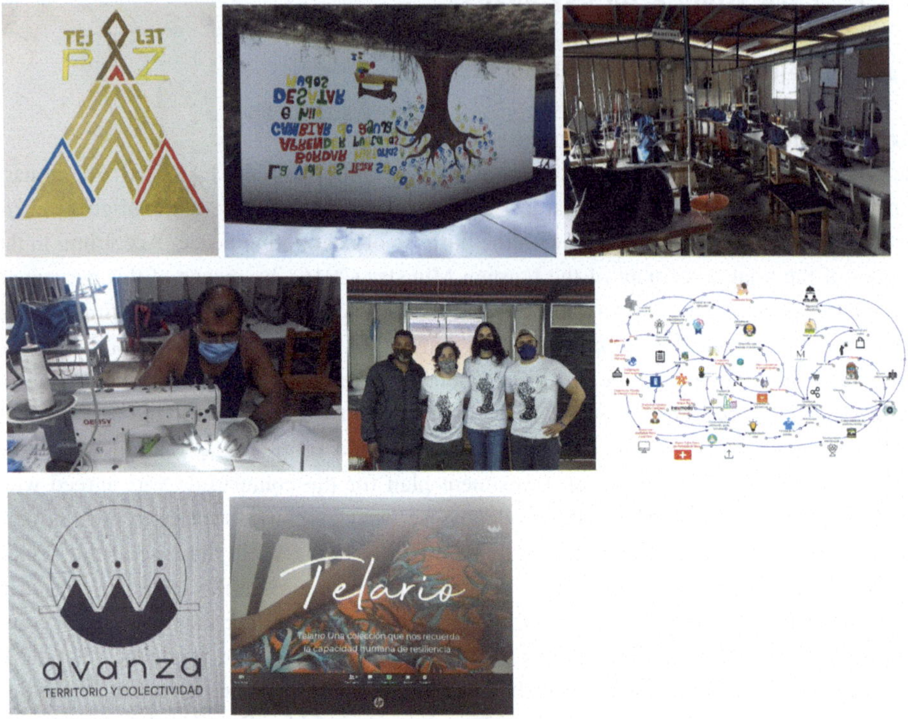

Fig. 4.2 Complexity Funnel Methodology and Social Transformation Workshops Peace Building Community 2 Int 2016 Industrial Engineering Students Camila Espinos y Valeria Rios Advicer Camilo Navarro University of los Andes Bogotá DC Colombia this figure

Tejiendo Paz Cooperative, as well as improve the quality of life of community members and support their reincorporation process. In addition, through this project it is possible to create a precedent for other organizations of this type, all with the aim of contributing to a lasting peace.

In this writing he documents the development process of the Social Transformation Workshops and how through a co-creation exercise with the Tejiendo Paz Cooperative, relevant conclusions could be reached. As a first approach, it is necessary to understand the reality of the organization and its context, the main actors, points of view and its objectives. This followed by an understanding of the operation, production process, and the relationship between the actors and the operation of the organization.

In a specific context, the Icononzo area is of high importance for the peace process, which is why a characterization of the situation from different approaches can be found in the literature. Karen Camila Rico (2020), in her research "Alliance between the Manifesta brand made in Colombia and the Tejiendo Paz cooperative, an initiative that bets on the reincorporation and the construction of peace in Icononzo, Tolima" talks about the textile cooperative peace, and the benefit of its alliances with the private sector. Specifically, the central objective is to systematize the experience of the alliance between the Manifiesta Made in Colombia brand and the Cooperativa Multiactiva del Común Tejiendo Paz as a strategy aimed at strengthening the reincorporation process and improving the living conditions of the population involved. From which she concludes that to promote this type of project, better training and formation of human capital is necessary, especially regarding organizational strategy issues. In addition, through the alliance with a private organization, a significant effort is made to improve the competitiveness of the cooperative.

The Tejiendo Paz Cooperative was born with the mission of contributing to the construction of a stable and lasting Peace, through the consolidation of productive processes. This, making use of all the resources and learning from the conditions of war, but now in terms of peace. On the other hand, they seek to use a sustainable and responsible development and production model; avoid extreme consumerism, waste, destructive relationship with nature, deepening the pattern of inequality, in order to promote principles of solidarity, cooperation, community and distributive and environmental justice (Our History 2018). In 2017, after the donation of a sewing machine by a professor from the National University, the cooperative's textile production project began; the Avanza brand. This brand launched its first collection called 'Telario' in 2020, made by 21 former members of the Farc-Ep in the former Territorial Training and Reincorporation Space (ETCR) Antonio Nariño, in the La Fila de Icononzo village, Tolima. This project has grown with the financial support granted by the Solidarity and Sustainable Territories project of the European Peace Fund, and the Multi-donor Fund for the Sustaining of Peace -MPTF-. In addition, the World Corporation for Women has done the accompaniment to make this company sustainable (Fashion that 'Advances' hand in hand with peace 2021).

At the beginning of this project, the Avanza brand had recently launched its Telario collection and had consolidated an alliance with the Inexmoda organization, which is in charge of offering personalized advice, as well as strong support during the training process. This with the purpose of developing the textile industry in the country and its level of competitiveness. At this time, the main challenge that the leaders of the organization considered was the positioning and visibility of the brand. On the other hand, there was a certain resistance towards believing in the impact that the project could have on the organization. Finally, it is important to note that this project was developed in the context of the Covid-19 health emergency, which meant having adapted to new working conditions and an atypical environment.

After the development of the 7 Social Transformation Workshops that make up the Complexity Funel Methodology, a better understanding of the organization was reached under a new paradigm in the world view. How it is spoken in the book "Web of Life" by Fritjof Capra all current problems cannot be understood in isolation. These are systemic issues, which means they are interconnected and interdependent. Understanding this means that society must leave behind the paradigm that it has dominated for many years, where the universe is understood as a mechanical system, composed of independent blocks, and change to a new paradigm. This new paradigm is characterized by representing a holistic view of the world, where the whole is evaluated as integrated parts instead of parts without association. The development of the workshops allowed us to have this vision of interconnected systems from which an action plan is proposed based specifically on the Systemic Archetypes, Pathologies, and Soft Systems where fundamental problems of the organization were identified. With this in mind, the final recommendation is to boost brand visibility through strategic alliances that also generate customer loyalty and attract new market segments.

4.5.2 Case Study 9 Emprenpaz

Case #8 EmprenPaz Peace Building Community 2

Emprenpaz

- Personas en Proceso de reincorporación de las FARC-EP
- Espacio Territorial Icononzo Tolima
- Cooperativa Emprenpaz.

4.5.3 Case Study 10 ASOMUFAN

Case # 13

Peace Building Community # 6 ASOMUFAN

ASOMUFAN

The “Asociación de Mujeres Farianas” Farian Women’s Association is a women association based in the Antonio Nariño ETCR in Icononzo Tolima Colombia. The Association was founded in 2009 and is formed by 25 women in reinsertion process. It seeks to provide a safe space and of overcoming women in process of reintegration into life civilian, mostly head mothers of family.

ASOMUFAN is one of the many associations that has a presence in the ETCR of the line at ICONONZO, formed in 2019 and currently has 25 active members, all women, most of them with children and without a sentimental partner. It was created under the idea of being able to provide a safe space where female Pharaohs could meet the needs linked to the process of reintegration into civil life for them and their families. Although legally they are a cooperative, the members state that they have not received the financial support agreed in the peace process. It is precisely that is why in their day to day they carry out various activities such as the sale of food or community events that allow them to raise funds economic for the development of larger projects that have as objective to create a space for professional and personal development that provides a steady income to support their families. Within your organization chart you can see two perfectly delimited areas. The first of these is the board of directors made up of Valentina, Viviana and Diana, on the other hand, is the assembly that involves the rest of the association. At this time, the association is looking for an injection economic given by an international entity for the development of the project more desired than it is to be able to have a plant for the production of creams and aloe-based cosmetics within the territory that provides them not only the economic income of which has been spoken but also a space of learning where they are the protagonists and can count on the knowledge necessary to produce, market and carry out all the activities necessary to keep a company and brand afloat emergent. Although so far only reference has been made to the projects that has the main impact on them as an association, it is important to emphasize that, throughout these years, they have been able to complete projects that have an impact on your entire community. Within them are the wellness centers that have been a great bet to the development and personal care of the inhabitants of the ETCR. They were also a great support in the construction of the playground that is a space where the little ones can have a healthy recreation with your friends and family.

Difficulties and tools

Throughout the project, there were some setbacks that were not they allowed the project to unfold as it was intended. One of them and the most important was the pandemic that negatively affected the possibility of having face-to-face spaces where interaction, communication and building relationships of trust are easier. The virtuality and implementation of communication channels such as Zoom, WhatsApp and phone calls, they were essential to be able to start conversations recognition of the community and advancement in the workshops. However, at throughout the process it was evident how the lack of contact made it difficult conversations and, therefore, the fluency of this project was selected for the interest in being able to provide the knowledge acquired throughout the career in a historical fact that it has great impact in various areas, mainly social. Peace in Colombia has been a work of several years that requires the support of all and to all those involved in it. This requires forgiveness and in this case in particular to be open to listen without judging the stories of the members of AMAZING that they are ex-combatants of the FARC-EP. This process of reintegration in which they find themselves is not easy and much less power start from scratch the development of various projects that have a positive impact on both the association and the community. This is why so it is interesting to know that the project will be able not only to be able to implement the Complexity Funnel Methodology, but rather it will also be possible to have a physical deliverable. On the other hand, the added value of this project lies in being able to help a community that requires tools to be able to belong back to civil life that has undergone several changes since they came out of this.

With the implementation of the complexity Funnel Methodology and the social transformation Workshops. It was possible to minimize the complexity of the social system by going from a coercive state to a pluralistic one.

The clearest achievement of the intervention in the projects was the decrease in entropy and economic collection to lead to carry out the painting day.

There are still conflicts and discrepancies that make it difficult to materialization of the cosmetic products project.

It is important to validate economic aspects in the short term, administrative and agrarian for the advancement of the production of aloe vera-based cosmetics.

Presence is considered essential to carry out the social transformation workshops and the application of the methodology of complexity funnel.

4.5.4 Case Study 11 Rugby for Peace

Case # 14

Peace Building Community # 6 ASOCIACIÓN DEPORTIVA ETCR Antonio Nariño Icononzo Tolima

We had the opportunity to bring rugby to the Antoño Nariño Territorial space in Icononzo Tolima. We run the transformation workshops together with the rugby for peace workshops to teach ex-combatants not only this new sport but also values, to give them tools for the peaceful resolution of conflicts. This allowed us to create safe spaces for dialogue between the different actors in the system, which contributed considerably to the reconstruction of the social fabric, generating new relationships, approaches and strengthening the community. In the case of rugby for peace, we apply methodologies developed by the University of Brighton in the United Kingdom to teach people in the process of reincorporation of the Antonio Nariño territorial space in Icononzo Tolima, not only the sport of rugby, which is a sport rich in values. but also to teach them strategies for conflict management and teamwork.

On different occasions we had the opportunity to go to the territorial space Antonio Nariño In Icononzo Tolima where we can publicize the sport of rugby, but also help with the socialization of conflict management strategies with people in the reincorporation process, helping them to generate peaceful dialogue scenarios where differences can be resolved by talking and reaching agreements.

4.5.5 Case Study 12 MAFAPO

Case #7 MAFAPO-Mamas falsos positivos de Soacha Peace Building Community 1

Mamas falsos positivos Soacha

The systematic disappearances of the men from Soacha and Bogotá, between January and August 2008, and the news that arrived days later about the discovery of the mass graves, put the women who today make up MAFAPO on an individual path to Norte de Santander in seeks to recover the bodies of their children, brothers, husbands and parents who had been falsely identified as guerrillas. "Falsos positivos" (false positives) is a term used in Colombia to refer to a grave human rights violation that occurred primarily between 2002 and 2008. It refers to cases where members of the Colombian military or security forces unlawfully executed civilians and then falsely presented them as guerrilla fighters or criminals in order to inflate their operational success rates.The practice of falsos positivos involved the abduction, killing, and subsequent manipulation of crime scenes to make it appear as if the victims were armed combatants who were lawfully killed during military operations. The objective was to artificially inflate the number of combat kills attributed to the military, thereby creating the impression of effective counterinsurgency efforts. The victims of falsos positivos were often poor, marginalized individuals, including farmers, young men, and individuals from disadvantaged communities. They were often lured under false pretenses with promises of employment or other opportunities, only to be executed and presented as guerrillas. The revelation of falsos positivos led to a major scandal and widespread condemnation both within Colombia and internationally. The Colombian government and military have since taken steps to investigate and hold accountable those responsible for these extrajudicial killings. Human rights organizations and victims' families have been actively seeking justice and reparations for the victims. Efforts to address the issue of falsos positivos have included reforms within the military justice system, increased scrutiny and oversight of military operations, and the creation of specialized units to investigate these cases. However, challenges remain in ensuring full accountability, reparations for victims, and preventing such human rights violations in the future.

They were isolated cases of women who were looking for an explanation for these atrocious events, and due to the ineffectiveness and lack of assistance from the state, legal medicine and judicial entities, they found answers in those who shared the same struggle. This is how the Association of Mothers of False Positives of Soacha and Bogotá MAFAPO was born "from the rivers of blood of Mario Montoya". From that moment, they became investigators, psychologists themselves and began to connect the dots that would lead them to the truth of the facts, but with all their efforts, most of their cases remained in impunity and although they were recognized by international entities, in the country they were not relevant, so much so that they came to be publicly branded liars. This situation changed for the better thanks to the peace accords. In the case of the mothers of the false positives from Soacha, we implemented the Complexity Funnel Methodology and the STW Social Transformation Workshops to support the Mafapo Foundation, which brought together the mothers of several children who were victims of false positives in Colombia and who worked with the national memory center for peace.

In 2016, the government of the former Defense Minister Juan Manuel Santos managed to sign the peace talks for the demobilization of the FARC-EP after six failed attempts in 34 years. With these emerged entities such as the Special Jurisdiction for Peace (JEP), the truth commission, and NGOs committed to clarifying the truth of the facts and guaranteeing non-repetition. The creation of these organizations has helped to advance in the cases of false positives such as with the sentence by the first specialized judge of Cundinamarca, in which 21 soldiers were sentenced for the cases of five young men from Soacha in March 2017; and the investigations carried out by the CPI since 2018 against 23 generals and six active and retired colonels for their involvement in 1,228 cases of false positives, including Mario Montoya, Juan Pablo Rodríguez Barragán and Jaime Lasprilla (Pacifista! 2018). Likewise, these organizations have opened spaces for the recognition of victims and the construction of memory, attracting attention and support from international organizations that promote the struggle that the women of MAFAPO began 12 years ago, giving them a voice and platforms to continue their resistance struggle against oblivion.

Engineering profession can contribute to unjust systems or be part of the systemic change towards social justice. During the last four years of peace in Colombia it has been possible to see the contribution of different areas of knowledge to the peacebuilding process, however, this discussion is traditionally exclusive to lawyers, political scientists, historians, psychologists and areas related to the social sciences. Engineering, however, is presumed to belong to a distant group that has no competence in this field. But environmental conditions in the twenty-first century have made engineers become more aware of their ethical responsibilities, roles in society, the consequences of their designs, interventions, management and above all the potential of the profession as one that can contribute. Greatly to change. This is precisely the case of Camilo Navarro, who asked himself the question: How to contribute to the construction of peace from industrial engineering?

Peace processes are highly complex because they involve a large number of actors and interests that often contradict and oppose each other; the attempt to reach a consensus makes them variable over time. Thus, industrial engineering becomes relevant, as it provides tools for organizational analysis and diagnosis, helping decision makers to recognize the changing nature of systems and understand the cause of certain problems caused by interactions between actors and their environment, as well. How to devise ways to solve them in the best way. Industrial engineering could also be said to immortalize these tools by taking them to the composition of replicable and adaptable methodologies to different contexts and spheres of peacebuilding in the national territory and abroad.

4.5.6 Case Study 13 Ecomisión Matamoros Proyecto Artemisa

Case # 9 Fundación Matamoros Peace Building Community 3

Matamoros Corporation-Corporación Matamoros:
The Matamoros Corporation (Founded in 1986), is a private non-profit entity that works for the benefit of the military and police officers wounded in combat and their families, as well as the widows and orphans of the fallen members of the Public Force in the line of duty.

It is an organization that is part of the Social and Business Defense Group (GSED) and seeks to complement the efforts of the State in caring for these people.

Project: Ecommission, where reforestation and reintegration converge.

This project is very special because they are two problems that go hand in hand. During the armed conflict the soldiers are affected psychologically due to all the traumas that they experience in the war, and as a consequence of this the ecosystem is destroyed at the same time. After both are affected by the conflict each takes a different path until eventually they come back together to restore each other. This situation is graphically illustrated below.

To achieve the social restoration of the victims, the corporation established clear objectives when taking into account in carrying out this project:

1. Environmental education: It seeks to train soldiers victims of the armed conflict with disabilities as environmental trainers in participatory ecological restoration. In addition to this, include the local communities where ecological restoration is carried out so that they preserve and continue the process with the soldiers. The above, in order that the communities that are those that reside in the territory most of the time have self-awareness and the capacities to maintain the restoration process in force over the years.
2. Participatory ecological restoration: It is sought through the national restoration plan (MADS, 2015) to restore one hundred hectares in strategic ecosystems, which will be strengthened with works of naturalistic engineering and erosion control. On the other hand, it seeks to implement research actions with detailed monitoring and evaluation. The above, in order to have empirical data that support the success of the program so that it can be taken to other areas of the country with greater ease and supported by real data.
3. Social inclusion for peace: It seeks to have a focus on the project in which, through the implementation of strategies focused on improving and reestablishing the psychosocial, family and community conditions of the soldiers who are victims of the armed conflict, they are reestablished in the community Colombian as proactive members of it.
4. Bio economy: Through this, it seeks to boost the local economy. On the other hand, this will generate incentives for the continuity of the project not only by soldiers who are victims of the project, but also by local communities since these are the ones that will last over time in the territories and must be taken into account for a successful environmental restoration.

In the first place, it is necessary to remember that at the beginning of the realization of the complexity Funel methodology, the system to be studied was in a conflictual coercive state. This is because the project was not being carried out. In the development of the social transformation workshops, together with the complexity Funel methodology, we were able to establish this system as a pluralistic one. The foregoing is due to the fact that we managed to go from a project merely written down on paper to creating a pilot plan in which we will see the development and implementation of what we have been developing throughout the document. Being a complex system, we cannot reach a unitary system since the pilot plan will be developed in the first semester of 2021, and it is there where we can robustly apply industrial engineering tools such as simulation models, resource optimization and improvements in the logistics of the project.

In addition, after identifying all the actors and variables of the Ecomisión project through the application of the social transformation workshops, Funel's complexity methodology and organizational cybernetics, we demonstrated that the Ecomisión project is viable. It should be noted that said viability was found in the diagnosis made through the "Viable Systems" workshop, where we identified that the project can easily be adapted and articulated to a viable system with the incorporation of actors that focus on specific roles so that the organization has independent viable systems and thus solve the organizational pathologies that arise at the time. These can be evidenced in greater depth in chapter five. Therefore, it is recommended that actors such as:

- General manager of the project.
- Operations room.

This will generate a System Four independent of System Five and also a System Three independent of System Five, since these functions are currently performed by System Five. This will ensure the strengthening of the systems and an improvement in the communication channels of the project. (check Staford Beer Viable Systems Methodology)

In the development of the analysis carried out, we were able to identify not only the viability of the project. In addition, through critical systems heuristics we were able to show which are the current gaps in the project and how the affected actors and variables can be involved in order to reduce these gaps. It is extremely important to highlight that the entire process carried out was validated and legitimized by the corresponding actors, ensuring that the project was a co-creation process in which a solution was found in conjunction with the actors involved.

To optimize the Ecomisión project we used the idealized design proposed by Ackoff in order to address the problem from a systems dynamics that was real to the current situation and to the changes that could be made in the project to generate the desired impact.

On the other hand, through the social transformation workshops and the complexity Funel methodology, they allowed us to reduce the conflicts and discrepancies that existed at the beginning (37) to a total of (12), which means a decrease of 67.57%.

In addition, it should be noted that our greatest value proposition emerged from these workshops and methodology, which is the realization of the pilot plan in which we can carry out the Ecommission project in the first instance and thus obtain empirical data that help us improve through the use of tools of industrial engineering to co-create a viable and lasting project. Through this proposal, it is intended that the Ecommission project be able to identify first, improvement opportunities for a project at the national level. Second, to be able to analyze whether said pilot plan produces the expected results with respect to the variable of interest of the project, which will be the number of victims restored annually. Third, obtain results from a pilot plan that will seek to attract investors and generate opportunities for growth and greater reestablishment of victims of the armed conflict in Colombia.

Although considerable progress was achieved, The understanding of the system by the decision maker was improved Organizational entropy shrank.

We must not lose the momentum that we have gained with the work of these months, so we set ourselves the task of finding 2 other students who could continue the process and continue with the other workshops to continue working hand in hand with Matamoros and his new allies.

Annex, we will be able to show the proposal made of the pilot plan in the El Zoque Natural Reserve.

We wanted to share with readers and those interested in knowing more about this pluralistic multi-methodology proposal the different case studies where critical systemic thinking methods and methodologies were applied to better understand complex social systems and help decision makers. The systemic application of these methods and methodologies in different social systems allowed us to identify which of these methods and methodologies were more easily implemented by different organizations and communities and generated more effective results in terms of complexity reduction, really helping in the decision-making process and optimization of resources, so we shared with you the different experiences we had during the last 6 years in the different phases of the project unraveling complexity in terms of the application of the Funes complexity methodology and the social transformation workshops. We hope that these cases are of great interest to readers and that they can learn from them to replicate the successful cases in other types of complex social systems that can sometimes be difficult to address.

Managing complexity in social complex systems is one, if not the greatest challenge of our time and we know that it is not always easy for the people in charge of these systems to know where to start or how to address the needs or problems that arise. That is ther reason why we hope that this methodological proposal will be of great interest and serve as support for all those people who are at the forefront of this type of systems where typical and classic interventions are not generating the expected results and require Finding a new way to face the new challenges, needs and problems offered by this new globalized world. and interconnected and increasingly complex, what we hope is that decision makers learn to see complexity as something good because as Neil Johnson mentioned, complexity is simple and if we are able to understand the elements of the systems, both the actors and the variables and their interrelationships, the boundaries, we will be able to understand their emerging properties and make a valuable contribution to understanding them helping the people who are in charge of them.

We hope that this methodological proposal is of interest to this group of people who are in charge of managing complexity within the systems they are in charge of, so that they can apply them and verify their effectiveness in terms of reducing discrepancies and conflicts within their systems, which undoubtedly helps the decision-making process, making the systems more effective, efficient and effective.

Reference

Navarro, C.A, and Dalrymple, O. (2019) Co-creation of a systemic model to support community engagement projects

Conclusion and Discussion

Thanks to the work carried out in 2020 and 2021 with the different peacebuilding communities, we realize that the implementation of the Complexity Funnel Methodology CFM and the Social Transformation Workshops STW allows effectively reducing the complexity within complex social systems by solving discrepancies and reaching agreements, generating a general understanding of the system and effectively supporting decision makers to make more informed decisions.

We observe that the gradual reduction of complexity is possible when we use the organizational entropy metric. However, it is very important that this is a continuous process and ensure that the projects with the peacebuilding communities have continuity over time. Failure to do so and in an anti-intuitive way would be creating the perfect bases for an even greater conflict as an archetype of load displacement (Braun 2002) by not providing a fundamental solution to the problem as an unwanted collateral effect, we may be creating the conditions perfect for a new conflict even worse than the previous one (Figs. A.1 and A.2).

The word Archetype comes from the Greek Archetypos, which means "*primitive or original model*". And these are considered the stepchildren of Systemic Thinking, the system archetypes were developed at Innovation Associates at the Massachusetts Institute of Technology or MIT in the mid-1980s. At that time the study of Systems Dynamics depended on the diagramming of complex causal diagrams and on the elaboration of computational models based on these diagrams only, using only quantitative tools, particularly mathematical equations to define the relationships between variables by calculating the partial integrals of the different levels to see its evolution over time. Charles Kiefer, the president of I. A., Suggested trying to communicate the concepts more simply. Jennifer Kemeny, with Michael Goodman and Peter Senge (2005) based in part on notes developed by John Sterman (2000) developed eight diagrams that help illustrate the most commonly seen behaviors. Some famous archetypes like "Growth Limits" and "Load Shifting" were translations of "generic structures", mechanisms that Jay Forrester (1961) and other pioneers of systemic thought such as Donnela Meadows (1972) described in the

C. A. Forero Navarro, *Untangling Complexity—Peace Building Engineering*, Synthesis Lectures on Engineers, Technology, & Society 29,
https://doi.org/10.1007/978-3-031-37038-0

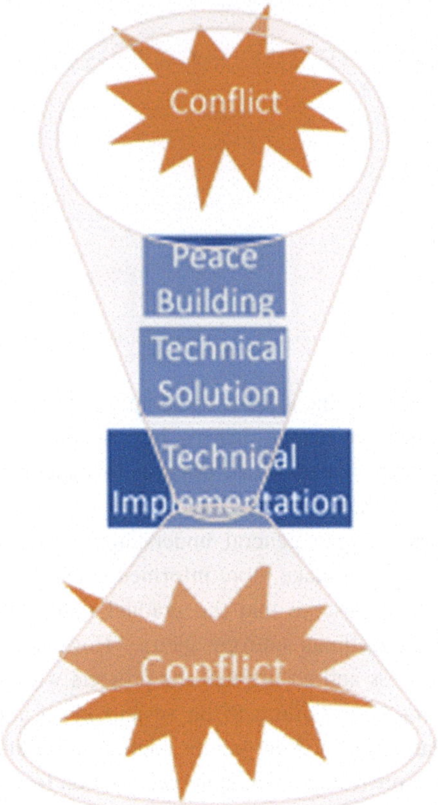

Fig. A.1 Bi-Funnel of unsuccessful engineering interventions in peace building communities creating an even worst scenario

1970s and 1970s. In the case of this Load Shifting Archetype, a short-term or symptomatic "solution" is used to correct a problem, with apparently positive immediate results. As this correction is used more and more, the fundamental corrective measures are applied less and less. Over time, the skills for the fundamental or ideal solution atrophy, creating greater dependence on the symptomatic solution, in the end an undesired side effect appears is that the symptomatic solution unintentionally increases the problem, which makes it increasingly difficult. give a fundamental or ideal solution to the Braun problem (Braun 2002). We then learned the importance of monitoring and control mechanisms in parallel to project execution.

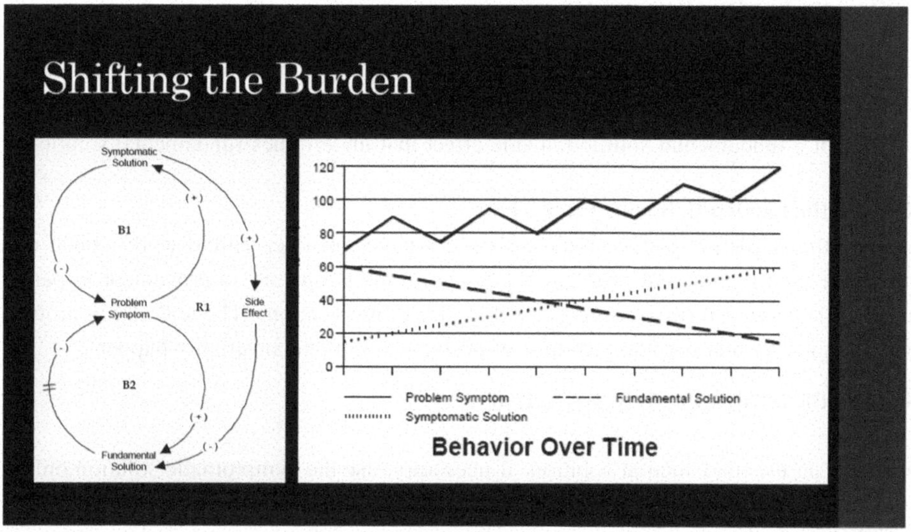

Fig. A.2 Shifting the burden system arquetype structure and behavior (Braun 2002)

Shifting the Burden is the first of several archetypes presented by Braun (2002) that illustrate the tension between

(1) The attraction (and relative ease and low cost) of devising symptomatic solutions to visible problems and
(2) The long-term impact of fundamental solutions aimed at underlying structures that are producing the pattern of behavior in the first place.

The tension between the two is understandable. Long-term solutions tend to demand deep understanding and learning about the underlying problem, take a long time to formulate, require a relatively large, up-front commitment of funds, and test managers' patience. All this in the face of pressures from many angles that demand that managers fix problems promptly and move on.

The essence of the Shifting the Burden Arquetype is that once the symptomatic solution (which by contrast requires less understanding, is easier to formulate, is relatively less expense-in the short run-and produces instant gratification) has had its effect, there is little perceived need to pay any more attention to the fundamental, underlying systemic problem. This has an unexpected result, a non desired collateral result that will increase the original problem and make even harder to give a fundamental solution to the problem.

Shifting the burden Dynamic Theory

This archetype states that a problem symptom can be resolved either by using a symptomatic solution or applying a fundamental solution. It hypothesizes that once a symptomatic solution is used, it alleviates the problem symptom and reduces pressure to implement a fundamental solution, a side effect that undermines fundamental solutions.

Shifting the burden Behavior Over Time

Shifting the Burden is one example of how management intervention works. Each time an intervention is aimed at problem symptoms, some temporary improvement in performance is experienced (this assumes a well-planned intervention). The underlying problem persists however and the reappearance of problem symptoms invariably happens.

Shifting the burden Prescriptive Action

- Focus on the fundamental solution. If necessary, use the symptomatic solution only to gain time while working on the fundamental solution.
- Elicit multiple viewpoints to differentiate between fundamental and symptomatic solutions and to gain consensus around an action plan.
- Use the archetype to explore potential side-effects of any proposed solution.

Shifting the burden-Seven Action Steps

- Identify the original problem symptom.
- Map all "quick fixes" that appear to be keeping the problem under control.
- Identify the impact of the symptomatic solutions on other parts of the system.
- Identify fundamental solutions. Develop multiple perspectives.
- Map side-effects of quick fixes that may be undermining the usability of the fundamental solution.
- Find interconnections to fundamental loops. Find links between the interaction effects and the fundamental solution that may be causing gridlock.
- Identify high-leverage actions from both perspectives.

What Does This Shifting the burden Archetype Behavior Really Mean?

Shifting the Burden is an example of creative tension at work.

The archetype draws attention to the gap between the pressures to perform in the short-term with the insights and long-term sustaining decisions to which systems managers seek to respond.

It also points to the critical importance of developing patience as one of the skills that systems managers include in their Personal Mastery of competencies. It illustrates the challenge and difficulty of demonstrating forward-thinking leadership in the face of mounting pressure to "fix it" and "get on to the next problem".

Without a clear and convincing picture in the manager's mind's eye (Personal Vision) as well as in the collective mind's eye of everyone (Shared Vision), the pressure to go for the quick fix may overwhelm the manager, condemning her/him to a recurring pattern of interventions that aim to solve the same set of problem symptoms (Braun 2002).

Relating these findings to the projects that we are promoting with the Peacebuilding communities in Colombia, we can say that, as the Shifting the burden Arquetype shows, if we are not capable as a state and as a nation to offer a fundamental solution to the problems of Peacebuilding communities, which is the search for a dignified life option that allows them to live well, through these community entrepreneurship projects we may be unintentionally creating the perfect setting for an even greater problem because Peacebuilding communities who have been working for years to carry out these projects as a real option to live away from the conflict if they realize or start believing that the state and the nation do not have a real interest in supporting them, they could once again take up arms against the state due to the bad conditions to which they may be subjecting.

This is perhaps one of the most important results of the work carried out over the years and really justifies the need to support these communities so that their projects become a reality and to be able to offer a real alternative to these people, ensuring long and lasting peace in the territory.

Preliminary Discussion:

This project has as its final objective a pluralistic methodology proposal to help peacebuilding communities in Colombia. Although the methodology as such has not yet been implemented in these communities, even though we are still in the planning stage of these interventions, we have learned a lot from previous experiences in both the private and public sectors and in working with communities. And we feel that both the TTS workshops and the CFM methodology are robust enough to take the next step.

We are confident that the execution, monitoring and control of TTS workshops in these peacebuilding communities can be done effectively.

We present in this document some of the applications of the TTS workshops and the EC methodology in different contexts in public mixed private companies and in work with communities. It is difficult to effectively assess the success or failure of the methods used with this new methodology proposal used during these 4 years, however the Organizational entropy metric is a very good proxy for how the use of this EC methodology and the TTS workshops are helping to reduce the complexity in the SSCs to facilitate the work of decision-makers another metric that we may be interested in reviewing is the number of relationships and joint work that result from the application of the workshops since, as we mentioned previously, this is a good proxy of the trust that exists within the system. What we can affirm is that since 2016 the workshops and the methodology have improved considerably due to the constant feedback obtained after each application. The application of the workshops has allowed creating and maintaining a cycle of continuous improvement where permanent monitoring of decisions and their impact on the system in terms of reducing social complexity measure as the number of obstructions both discrepancies and conflicts, and how these have been solved by means of the creation of safe spaces for dialogue and the creation of agreements between the different actors, which has given legitimacy and continuity to the projects, acting as a catalyst for them and ensuring that they are always aligned towards the common objective of the actors.

We learned about the importance of monitoring and control in parallel to the execution of projects in the case of failure with the mayor of Cartagena and we transmitted these experiences to decision makers in other SSCs. This so that they understand the importance of identifying something that is not going as planned using specific metrics related to the variables of interest. This allows them to be informed in a timely manner of the anomalies and to take immediate control actions to reevaluate the intervention that they are making in the systems. SSCs are dynamic and non-static systems, they change during interventions and due to their changing nature, the systems must have the characteristic and be adaptable to these possible changes and decision makers know how to do it. In the case of the TTS and the CFM The changes that were made in the systems have always been reflective and are directly supported by the results of the application of the workshops.

However, we saw examples of success and examples of failure. The implementation of the workshops with the mayor of Cartagena was an example of failure because after the intervention we were unable to ensure the continuity of the project and especially the subsequent monitoring and control part. After achieving the expected results, that is, after achieving the implementation of the program and its execution, as the preliminary results were favorable, the motivation to continue monitoring and control was lost, and after a time without this monitoring, the system not only failed but I get to be worse than I was before the intervention causing the resources and time dedicated to the project to be lost. In all other interventions, the cyclical use of systemic thinking methods has contributed to the continuity and success of the projects that were co-created and implemented. The new methodology, however, showed certain limitations because, although the implementation was successful, we have problems with the monitoring and control part, especially in the post-implementation stages. For subsequent implementations of TTS and EC methodology, we hope that we can improve these stages of post-intervention follow-up.

The following interventions foreseen for the CFM are intended to be carried out in peacebuilding communities in Colombia in former Veredales or Territorial Spaces and, if successful, we would like it to be possible to address other larger systemic phenomena in Peace Building in the country, but For this, it would be necessary to expand the limits of the intervention and include higher levels of leadership, and the relationship with the creation of public policies without taking sides or favoring any political actor since any position could inevitably alienate certain actors who are part of these systems. Make our project an unacceptable project and losing all legitimacy and continuity. However, as Foucault [55] mentions in his work The Archeology of Knowledge, being apolitical is often not an option because, as he mentions in his work, there is always a close relationship between knowledge and politics and they do that depending on the The government's agenda in turn, the academy is aligned to certain objectives and the projects that arise from it are political, even if the organizers and the participants do not particularly identify them as such. For this, it is necessary to improve the methods used in the TTS workshops and ensure their replicability through training and access to more regions. There are not many works or methodologies that address the Coercive or Complex SSC intervention, which requires more rigorous work and a more comprehensive approach that encompasses both individuals and their way of thinking and their diverse points of view. Despite presenting obvious limitations, this project makes important contributions regarding the role of engineering in working with communities and mainly in working with peacebuilding communities. We show how the use of the new methodology and workshops positively affect SSC systems, reducing their complexity and facilitating decision making. It also contributes to the need mentioned by Howick and Ackerman and later by other authors such as Henao and Franco to enrich the academic dialogue of projects where multimethodological interventions are used.

Our work shows new ways of applying methods and tools such as critical systems heuristics, soft systems methodology, interactive planning, organizational cybernetics, organizational pathologies, systemic archetypes, among other tools. According to Jackson qualitative tools for dialogue and agreement creation, typical of the social sciences, have not been used for conflict management and resolution before 2006, so if there are new proposals, they are no more than 13 years old and have not yet Because they are so well known, we contributed to broadening our knowledge of these multimethodological interventions and their results, with special emphasis on the obstructions mentioned by Ackoff (1993) in the discrepancies and conflicts defined in his book Idealized Design and addressed thanks to Interactive Planning. The application of STW has taught us a few things about safe spaces for dialogue and the creation of agreements between the parties involved is still not a perfect methodology and has much room for improvement.

The success of this methodology is mainly based on the creation of safe spaces for dialogue and the co-creation of agreements between actors, ensuring safety and respect during them and helping the empowerment and active participation of all actors to expose their views, roles and interests and to adapt to the system. Power relations limit our ability to express our thoughts and therefore limit our actions.

STW help actors to overcome these power relations and speak freely to improve understanding of their role in the system, their motivations and interests, the Workshops then have an emancipatory role in the process and also allow all actors to recognize the motivations, roles and interests of the other actors in the system. They allow actors to have a safe space to do catharsis and talk about their disagreements, develop the ability to critically reason about their actions, their decisions and how these affect others, allowing them to actively participate in a critical and constructive discussion to better adapt. to the system to which they belong, also transforming the power relations that exist within the system. The new EC methodology and the TTS workshops a new perspective on how to use engineering tools to support work with organizations or communities when there is a concern about the freedom of the actors due to power relations mainly in the Coercive or Conflictual SSCs (Jackson and Keys 1984).

Our research also shows that developing stakeholder empowerment in SSC and their active participation makes more actors want to be part of TTS by improving the overall understanding of the system. Research is showing that the use of empirical-analytical quantitative methods fosters not only the empowerment of the actors that are part of the system but also the appropriation and legitimization of the solutions that are co-created by the system's actors, which helps to ensure the continuity of projects and the implementation of agreements, measures and solutions. From the previous implementations of the TTS we have learned how to generate safe spaces for conversation, giving voice to all the actors, generating a space for reflection and co-creation. Another important contribution of the EC and TTS methodology is that it shows the positive impact of adopting a systemic approach when working with communities seeking improvement of the system as a whole and not its parts as proposed by Capra (Capra and Sempau 1998) in La Trama De La Lifetime. And from the importance of framing well the SSC system we want to work with using the boundary limits of Ulrich's Critical Systems Heuristics (), we also learned that TTS can be addressed using different methods or different parts of methods in a relationship synergistic with each other. The TTS were adapted according to the feedback of the different actors of the different complex social systems with which we work and in the same way our understanding of the communities with which we work and the way in which we should interact with them. Framing the Social System Studying is very important with what Churchman (Vhurchman 1957, 1970, 1974) Ulrich (Ulrich 1987) and Midgley (Midgley G 1992, 1995, 1997, 1998, 2011, 2013) mention. Criticizing these boundary limits can help researchers make these critiques in a more thoughtful and dynamic way and not just take limits for granted because this is a process of continuous improvement. This considerably improves our understanding of the actions, facts, feelings, motivations, and value systems of the various actors that we must consider, as well as those that we do not consider are so relevant and can be neglected. The border limit trials that we managed to establish together with the actors considerably affected our work as we realized the changing nature and the need for rapid adaptation in these dynamic systems since the modifications within the system change the border limits and when changing Boundary limits change the things we consider most relevant and these changes affect the system as a whole.

Finally we want to mention that the use of this new methodology and the TTS that are described in a limited way in this document illustrate in a very clear way the transformative potential that the use of these engineering tools can have in working with communities, which shows very The work that we want to carry out with the peacebuilding communities is promising, which is the new step in our project. The EC methodology and TTS workshops can be seen as a possible option when working with peacebuilding communities and can be used to compare the success or failure of interventions in these communities. The workshops seek to implement different strategies to give voice to the actors who are further down the chain of power relations so that their perspectives, points of view and values help reflect on decision-makers, guaranteeing the proper development of projects and their continuity thanks to the fact that they involve the largest number of actors in the system in the co-creation of solutions. In our case, the success of the programs always exceeds the initial group of actors with which to start because when seeing the potential of the workshops and the new methodology, good practices and co-created solutions are extended to larger groups and more extensive as seen, for example, in the work carried out by the ASF Advancing Students Forward foundation in both Tijuana and Sandiego.

The application of the social transformation workshops to the peacebuilding communities is intended to begin in January 2020, possibly in one of the FARC-EP's territorial spaces in the municipality of Icononzo Tolima.

Conclusion, discussion and recommendations

Preliminary Conclusions

By way of preliminary conclusions, and to end this chapter of results, we have prepared a methodological proposal to support decision-makers in complex social systems, focused mainly on Peacebuilding Communities in Colombia and the projects that derive from them. Projects that seek to offer real alternatives for employment and development for these communities.

In the first place, it is an *eminently dynamic and constantly changing* methodological proposal. The main variables that are presented as substantial for the understanding of the Complex Social Systems (CSS) studied show complex dynamics, whose patterns of change are not easy to study and describe. That is why we should not try to analyze those systems or propose interventions as they were static. The dynamic characteristic of those systems must always be taken into account so, we are flexible enough to adapt to appearing changes.

Second, the methodological proposal of Complexity Funnel (CFM) and the Social Transformation Workshops (STW) have shown *Complex Dynamics* when the key variables of these systems are intervened, presenting *chaotic dynamics*. With this condition, the model offers a holistic view of the state of these dynamic systems that are credible and legitimized by the actors that make up the social system. This makes this representation of the system an important visual tool to be able to align the actors towards common projects, thus gradually overcoming their differences and working together for a common objective. When the change in key variables due to the resolution of obstructions, both conflicts or discrepancies, is effective, we observe as a direct and indirect consequence a behavior in other secondary and tertiary variables of the model (2nd 3rd-degree effects), such as the cancellation of any of them or the exorbitant growth of the same.

Thirdly, it has been verified that, indeed, the dynamics followed by the proposed complex model is chaotic and permanently changing, but through the application of the different Social transformation workshops, it has been possible to identify the high points that deserve greater attention from the Decision-Makers, decisions regarding the resolution of possible discrepancies and conflicts through dialogue and the creation of agreements.

Projects that we are promoting with the Peacebuilding communities in Colombia, are delicate as the "Shifting the burden" archetype shows in our last chapter, if we are not capable as a state and as a nation to offer a fundamental solution to the problems of Peacebuilding communities, which is the search for a dignified life option that allows them to live well, through these community entrepreneurship projects. We may be unintentionally creating the perfect setting for an even greater problem in the Future, a greater armed conflict, because Peacebuilding communities who have been working for years to carry out these projects as a real option to live away from the conflict if they realize or start believing that the state and the nation do not have a real interest in supporting them, they could once again take up arms against the state due to the bad conditions to which they may be subjecting.

Discussion and Conclusions

There are many conclusions that we must stop at in this last chapter of this book. Most of them will refer to the consideration that we have established as our research question that if it is possible from engineering to contribute significantly to facilitate the decision-making process in complex social systems and effectively promote own projects, giving them legitimacy on the part of the actors and continuity in time by using critical systems thinking and complexity management to support community development.

Throughout these 6 years of work, we have been able to demonstrate that engineering can effectively help decision-makers of complex social systems to make better decisions, decisions that allow accelerating the materialization and implementation of projects, ensuring their continuity and their legitimacy over time.

It was possible to demonstrate the relevance of some engineering tools mainly, Critical Systems Thinking Method's and Methodologies, Project Management and Participatory Action Research and their combination, and how they can be used successfully so that complex social system interventions can effectively reduce the complexity in these systems and help those who have the difficult task of making decisions that They will affect many people based on incomplete information and limited resources so that the projects that arise from these decisions are carried out to fruition.

It was possible to demonstrate the importance of visual tools for the holistic representation of complex systems and the importance of dialogue in the process of identifying obstructions, both discrepancies and conflicts, and the importance of solving existing discrepancies by creating common mental models and allowing that, in the event of conflicts, mid-point agreements can be reached that allow continuity and effective monitoring of the projects while these actions are legitimized by the actors.

We also demonstrate the importance of supporting these projects and ensuring that they are successful to offer a real alternative to peacebuilding communities. We show that if these projects are unintentionally unsuccessful, they can create the perfect conditions for further conflict as an unwanted side effect. It is in the interest of the state of the nation of all Colombians that these communities are fruitful in their endeavors and can really use these projects as the necessary impetus to change their reality to create opportunities to reduce inequality gaps.

Complex social systems are *dynamic systems* in constant change and due to the obstructions, that exist within them, they are systems far from equilibrium, such as chaotic systems. On the other hand, we will also dwell on some conclusions derived from the Complexity Funnel Methodological proposal and the Social Transformation Workshops. To do this, at first we will discuss the consideration of complex social systems seen as *dynamic models* or systems far from equilibrium and complex systems, due to the existence of obstructions, both discrepancies and conflicts, and in a second place, we will talk about the reduction of this complexity due to the application of the social transformation workshops and how these workshops allow to accelerate the processes of implementation and materialization of the projects, giving them legitimacy and continuity over time.

Through the development of the Organizational Entropy (OE) metric, we can then clearly and importantly monitor the systematic reduction of complexity in complex intervened social systems and then know if our interventions are really helping decision-makers.

Regarding the main objective, this work has defended the main thesis;

> "*That from engineering we can help decision makers in complex social systems to make better decisions, with a clear focus on peacebuilding communities in Colombia.*"
>
> (Navarro 2022)

When we speak of better decisions, we refer to decisions that provide solutions to existing problems or needs but that are legitimized by the actors and that have continuity over time. The social systems studied among which the Peacebuilding Communities are complex systems and one of the main originating sources of this complexity is the chaos that reigns within these systems. Peacebuilding communities can be understood as chaotic systems out of balance. Disorder or chaos is the most economic and natural operation for any complex social system that seeks survival among its ends (Navarro 2001). Being able to understand this complexity and reduce it in order to make better decisions regarding the use of limited resources with incomplete information is that the success of the projects that are born from these systems.

The predominant paradigm in systems theory and organizational theory tells us that

> …systems consciously adapt to their ends and needs
>
> (Peiró 1990).

This paradigm has overemphasized the concepts of balance and stability, coming to propose casuistry about which are the most effective organizational structures and systems depending on the environment in which the system or organization moves, this environment being external and alien to the systems themselves or the organization itself as explained by Navarro (Navarro 2001) (See 1959).

This paradigm begins to falter due to the increasing influence of different fronts and points of view. The academic discussion has still been alive since the 1940s and the great crisis that the science of complexity experienced in the 1990s. One of the widely studied is the application of mixed and multi-methodologies that some call pluralistic, a combination of different methods to intervene in complex social systems. Here we highlight the works and academic reflections of authors such as Habermas, Jackson, Ulrich, and Midgley, and others. Another paradigm comes from the new proposals and the new systemic developments and arises when in the literature of systems and organizations and their management the topic of disorder, chaos and turbulence appear more and more frequently.

Chaos, disorder, systems far from their equilibrium states, turbulence, which are properties of the environment, begin to manifest themselves more and more as generic aspects of current environments and also of social systems and organizations. And that is why traditional and rational models of systems or organizations fail to successfully explain the reason why social systems are involved in dynamic processes of permanent change, sometimes even radical changes. As Weick says:

> ... instead of stability, change is the rule of any complex system...
>
> (Weik 1969)

Chaos, disorder, turbulence, uncertainty seem to be endemic to social systems. And that is why this work emphasizes so much the need for monitoring and control in order to manage complexity. The raw material with which a professional works when he seeks to make an intervention of a social system is information that he seeks to translate into data. However, the problem is that these news programs are ambiguous, subjective, uncertain, and confusing, rarely do the professionals in charge of these interventions work with absolute certainty of what is going to happen in the intervened system.

It is true that as human beings, we are animals of habits, and as such, we do not have a high tolerance for uncertainty, we choose as heads of the systems and as decision-makers people who make us feel safe, who sell us an illusion of control, stability. The lack of stability control of understanding generates anguish given our inherent need to control our environment. Human beings are creative and ingenious by nature, and we need to be able to experience change, novelty, have the experiences of surprise. Humanistic psychologists such as Moreno and González were able to see very well how creative spontaneity is a fundamental capacity of every human being, even related to the notion of health (González 1987).

If we then see social systems as complex systems, order and stability cease to be a goal to be achieved by decision-makers, and in the same way, disorder and chaos cease to be undesirable states. Complex social systems are systems that will never reach those states of equilibrium desired and promised by rational models. On the other hand, complex social systems are in a permanent struggle for survival by the adaptation for continuity and are permanently involved in processes of change, suffer instabilities, and in the small changes due to interventions that seek to reduce complexity often introduce important behavior news. Complex social systems find their best way of functioning when they manage to be flexible adaptive and manage to function in an emergent and spontaneous way, understanding the changing realities of their environment, being able to socialize them through dialogue, being able to identify the obstructions to the legitimacy and continuity of projects over time and offering real solutions to them.

Complex social systems are systems that will never reach but do not seek to reach a state of equilibrium either, these systems are characterized by their inherent instability due to their ever-changing dynamics and by the presence of self-organizing phenomena. In complex social systems, we observe processes non-linear in nature and in which chance plays relevant role in determining their evolution, in complex social systems it is impossible to really predict their future state with certainty. It is precise because of all these characteristics that complex social systems that find their origin in the fundamental need and will to organize and align themselves towards common goals becomes difficult. Chaos, disorder, distance from equilibrium, non-linearity, permanent instability, dynamism, chance, the impossibility of predicting the future, the existence of paradoxes, are sources of the ambiguity necessary to start the task of organizing and making the right decisions regarding the management of these systems (Navarro 2001).

> "... *Organizing is coordinating interdependent actions within perceptible sequences that generate good results.*"
>
> (Weik 1969).

The organization of the actors towards a common objective, aligning themselves to carry out projects that lead to that expected result is only possible thanks to the disorder and chaos of the previous state. But, in addition, although the task of organizing a complex social system and helping to make good decisions consisted in reducing this ambiguity, the key is not to eliminate it but to manage it well. Deep down, complex social systems have an explicit recognition that ambiguity and uncertainty are necessary for their proper functioning. Disorder and chaos are essential for the gradual appearance of order, for the resolution of discrepancies for the creation of agreements that allow conflicts to be overcome and that is why the convenience of introducing ambiguity to generate spaces for dialogue is even proposed. Managing, making decisions, organizing consists according to Weick in

> "... *An ambiguity or complexity reducing process.*"
>
> (Weik 1969)

Other authors such as Navarro (2001), Peiró (1990) and Jackson (2019), are also aligned with these ideas of Weik and Navarro here appropriate to propose an approach to the study of information processing in complex social systems and organizations follow this idea not only of complexity reduction (Ambiguity) but of ambiguity management or complexity management. As Navarro mentions;

> "*If the environment in which the complex social system moves is chaotic, the processes must also be chaotic for a better system-environment adjustment.*"
>
> (Navarro 2001)

Social systems are becoming more and more complex (Johnson 2007) the social systems of the peacebuilding communities studied are no exception. These complex social systems also become complicated and coercive, making it more difficult to intervene successfully.

In the systems theory literature, as in the organizational and management literature, the complexity of social systems and organizations is mentioned. But, in the rational paradigm of systems theory, the complexity of systems is often reduced to a quantitative aspect. A good example of this is mentioned by Peiro and Navarro.

> "*Complexity is usually referred to ... "to the multitude of structural units in which the members of an organization are grouped.*""
>
> (Peiró 1990; Navarro 2001)

It is also common that they refer to environmental complexity as;

> "... *Dimension of the organizational environment and understood as the number of relevant factors existing in that environment in which a system operates.*"
>
> (Child 1972).

> "*Complexity has little to do with number and a lot to do with relationship.*"
>
> (Navarro 2001)

This discussion is addressed in detail in part 7 of this chapter when we talk about complexity and the arguments of Habermas, Jackson de Ulrich. Midgley propose that complexity has to be understood as an interaction between 4 types of complexity, the first 3 aligned to the 3 worlds proposed by Habermas and the 4 the ontology or the interaction between these 3 worlds.

If we think, for example, of the infinitely complex dynamics that an equation generates in a dynamic system and how with the change in a single variable its behavior sets a complete imbalance of the system we see how little variations in a social complex system can actually have strong effects. Another very prominent aspect of complex social systems is their explicit recognition of the importance of non-linearity, also recognizing the limitations of our linear thinking and not trying to adjust it to it but recognizing that the world is not like that, that the world instead from linear is divergent it is interconnected and that these interrelations become more and more complex. To understand the effects of non-linear processes, it introduces the temporal factor, in the manner of continuous iterations, and makes use of accumulated experience and the possibility of building models and simulating them to glimpse possible dynamic developments in any organizational process. We have shown that nonlinearity introduces radically new qualitative behaviors. Now we have arguments to defend those nonlinear processes that cannot be studied from linear approaches. Management and decision-making based on linear reasoning stumble upon the complexity of non-linearity and becomes useless as a management or decision-making tool in complex systems. To manage complex systems and to have a good decision-making process we need to be able to understand and manage complexity.

Throughout these years, it was also evident that the relationship that complex social systems have with their environment is much more complex than what the rational paradigm has shown us. Social systems adapt to their environments, submitting to the conditions that they impose on them, this is the only way to survive. Precisely to avoid this, and as a complement to the adaptation process, complex social systems self-organize by rebelling against environmental conditions, a process that is clearly related to that of operational closure proposed by the theory of autopoietic systems (Varela and Maturana 1973) According to Maturana and Varela, systems that present a network of processes or operations (that define them as such and make them distinguishable from other systems) are autopoietic, and that can create or destroy elements of the same system, in response to disturbances in the environment. Otherwise, that the evolution of systems consists of a continuous succession of periods "of adaptation and periods of self-organization" that are intermingled, the latter being the resource so that small organizational fluctuations are amplified and are able to favor new self-processes. -organizational that give rise to new forms of organization. We spoke in the first chapters of the concept of community resilience and how we can relate this concept to the search to return to a more organized previous state or to seek a more organized future state, in both cases, the communities seek to reduce the complexity of the systems of the communities they are a part of.

For the interventions carried out to lead social systems to new states, misnamed "states of order", certain isolation from the environment is required since it always tends to cushion and neutralize the interventions.

With the double evolutionary process of **adaptation and self-organization**, the relationships between environment and systems become more complex. In the development of these Peace-Building community base projects, adaptation and self-organization processes intervene.

On the one hand, the environment influences systems and requires an adaptive process for their survival. It is important to mention that these adaptive processes often occur empirically out of necessity, but they do not follow a clear process. Systems are capable of self-sustaining within their own limits, internally delimiting their caps and influencing the environment itself, thus showing a capacity to co-evolve with the organizational environment. With this process, the systems can modify their environment to a certain extent to make it more conducive to their interests. Wegner Ulrich tells us about the importance of border limits and about being able to understand very well what are the motivations, the power relations, the expertise, and the legitimacy of the systems in which we work (Ulrich 1987).

It has been repeatedly argued in systems theory that changes in systems are caused by the responses that they give, or are to give, to changes in the environment. The sciences of complexity trigger internal mechanisms in the systems that are the source of change and, in turn, internal system changes are capable of influencing the environment itself. Disorder, chaos, and complexity make systems self-organize by rebelling against environmental conditions. Autopoietic systems, living systems are closed systems that only refer to themselves in their interactions with the environment. This second idea contributed by the theory of autopoietic systems is of great relevance, and they come to strongly influence the importance of the internal dynamics of complex social systems, in the closed nature of management and organizational decision-making processes, it should not be confused with the fact that the system remains open to outside information. Autopoietic systems emphasize the importance of internal processes, with which the long-held theme of disorder, chaos, and turbulence in the environment acquire another reading. The turbulence is not only in the environment, it is within the organizations that make up that environment.

Environments, in some way can also be created by the same systems and they begin to have an effect on them once the processes are quite advanced.

The main reason for the complexity of the systems lies in a new thermodynamic principle that, as we have also said, is not recognized by official science, but it is there.

> "*Social systems, like other open and complex systems, show increasing evolutions in complexity and internal structuring.*"

Somehow, it seems as if living for an open system has only one possible direction. Either it takes the direction of progressive complexity or it regresses until it becomes a closed system and, consequently, a dead system or destined to die from its inability to adapt.

Once complex social systems have been conceived as self-organizing systems and that said property is natural and advantageous with a view to a better evolution concerning the environment, it then becomes clear that a way must be found on how to manage these complex social systems. Self-organizing and help the people in charge of decision-makers to make better decisions, decisions that allow continuity and legitimacy to the projects over time. We then asked ourselves how we could facilitate this self-organizing capacity that generates a greater understanding of the system that reduces complexity and that allows accelerating the processes of implementation and materialization of projects. It is at this time that we present our methodological proposal for Funnel of complexity and social transformation workshops that seeks to offer guidelines for possible interventions of communities identified as complex social systems.

The social transformation workshops (STW) have a clear socio-technical focus and use methods from different methodologies of critical systems thinking of organizational management and interactive planning. They are design guides compatible with a complex conception of social systems.

The implementation of the workshops and their results are of greater interest in part because of how novel they are in the eyes of the traditional paradigm and, at the same time, how realistic and practical their results are in terms of the decision-making process and day-to-day management of these systems. The 12 workshops suggested to decision-makers have been worked by numerous decision-makers in closer settings. There is an important change in the way of seeing the system no longer as something that I can control but as something that I need to manage and for this I must be able to make correct decisions that allow the survival of the system. The idea of decision-making and management is definitely changing as actions that seek to control the system.

Along with abandoning the pretense of control, decision-making and complex management recognize that the long-term future is unpredictable, that it is difficult to attribute an observed effect to a specific cause or sets of causes, that performance irregularities are a fundamental property of the system.

As Stacey would say;

> "... *To operate in the new frontier of disorder and chaos, managers, in their strategy, must prioritize the understanding of the systems as an interconnected whole, and not operate and understand the isolated parts of it; they must be anchored in the here and now, instead of temporarily positioning themselves in the future; and focus on processes rather than results.*"
>
> (Stacey 1992)

One aspect that could be criticized both to the design guides and to the complex management principles that have been proposed is the absence of limits in them. It is true, we have agreed on the importance of creating uncertainty, enhancing positive conflict, etc., but to what degree? In the absence of research in this regard, one can only use common sense and be guided by the maxim of achieving and maintaining the best quaitiy life levels for the organization members.

From these conclusions, discussions and reflections, some questions also arise.

Are all complex social systems (SCS) disordered, chaotic or turbulent?

Based on the experience of recent years, we could affirm that yes, regardless of the type of mixed public–private system, the national or international community in all the systems intervened in these years, we have found the characteristics that show us the high complexity that reigns within these systems and how difficult it is to exercise an appropriate decision-making and management process that allows a cycle of continuous improvement, giving continuity and legitimacy to the projects over time. It seems as if the proposals made in this work are more congruent with systems of a certain type due to the approach we make with the Peacebuilding communities. We could affirm that disorder, chaos, the state away from equilibrium is almost the permanent state of these systems, almost like a continuous state, with a large spectrum from less to greater chaos, disorder, ambiguity, or complexity, thus, in large companies, For example, moments of chaos or disorder are minor and almost punctual between large periods of stability and that is why we observe in their behavior a much more effective reduction in complexity due to the application of the workshops. On the contrary, in the most bureaucratic public sector companies and in the communities both nationally and internationally, and mainly in the different peacebuilding communities, moments of disorder and chaos are more prevalent than moments of stability.

Conclusions

To conclude, we will return to the original questions and initial hypotheses presented at the beginning of this work and the conclusions we have reached in relation to each of these. These hypotheses have been presented in the first chapter.

Regarding the question of "What responsibilities and beliefs do engineers hold when exercising our profession?" more and more ethical discussions about the transformative role of engineers in society and the responsibilities that this entails are brought to the classroom. We have to continue promoting these topics and these conversations so that our future engineers are integral professionals not only people with technical scientific knowledge but also with clear ethical and social values. Regarding the question "How are engineers educated to solve local and global problems with peace & social justice in mind?" We review the particular cases of various academic programs that are managing to fulfill this task; however, they are still very few and we are far from being the standard. It is important to keep talking and sharing the importance of including these vitally important issues in the training of engineers.

1. In the first place, and in relation to the first hypothesis indicated in Chap. 1 that;

"Complex social systems are open systems far from equilibrium"

We have shown how a complex, disorderly, chaotic, far from equilibrium model manages to give a better account of the task of organizing in terms of generating meanings. From this perspective, the ups and downs and the permanent changes in which social systems are involved are not surprising. From this perspective, the importance of uncertainty is understood and recognized as an element to be managed, not to be absorbed.

2. In relation to our second working hypothesis;

"They are Industrial Engineering tools methods and methodologies that can support Peace Building processes"

Industrial Engineering (IE) tools, methods, and methodologies can be applied to support peacebuilding processes. While Industrial Engineering traditionally focuses on optimizing systems and processes in manufacturing and operations, its principles and techniques can be adapted and utilized in various contexts, including peacebuilding initiatives. Here are a few examples: Systems Thinking: Industrial Engineering emphasizes systems thinking, which involves understanding the interdependencies and interactions within a complex system. Applying systems thinking to peacebuilding can help identify the root causes of conflicts, analyze the dynamics between different actors and factors, and develop holistic strategies for sustainable peace. We have proved how engineering tools methods and methodologies can successfully support Peacebuilding Community based projects accelerating the implementation and materialization of the projects acting as a catalyzer. They also help give continuity to ongoing projects and legitimacy from all the actors that are part of the social complex systems intervened. Applying Industrial Engineering tools to peacebuilding requires a contextual understanding of the specific conflict dynamics, cultural factors, and stakeholder engagement. Collaboration with experts in peacebuilding, conflict resolution, and local stakeholders is crucial to ensure the relevance and effectiveness of these tools and methodologies in supporting peacebuilding processes.

3. In relation to our third working hypothesis;

"In the development of these Peace-Building community base projects Adaptation and self-organization processes intervene in the development of all complex social systems"

We have shown how an adaptive organizational development model is not capable of accounting for organizational change itself. We have shown that systems, in addition to adapting to their environments, are capable of innovating in them and creating new environments. This capacity for innovation is a clear example of how complex social systems not only meet the requirements that their environments demand, but are also capable of closing in on themselves and self-organizing as an elementary condition for the production of the development of innovation processes.

4. - Finally, in relation to the fourth hypothesis proposed in the first chapter that;

"The complexity that we observe in many of the psychosocial processes that we study within systems can be due to very simple causes"

Simple models of holistic representation of the social systems intervened demonstrate to be capable of producing better understandings of the systems, supporting the decision-making and management processes within the systems, accelerating the processes of implementation and materialization of the projects proposed by the communities. According to this the complexity can be the result of simple causes. Delving a little deeper we have also shown that:

The dynamic process, interventions to support peacebuilding community projects can be studied and supported from the paradigm of complexity, in such a way that we have developed a methodological proposal that allows supporting these interventions. And this is based on a classical approach in the study of the complex social systems intervened (studying them through the use of classical methods such as questionnaires, surveys but also generating spaces for safe dialogue and giving voice to all the actors) and setting a methodological proposal that let us verify if it shows a plausible behavior with the relationships found empirically.

In relation to this, we have verified that the dynamics of the studied social complex systems are chaotic and sensitive to changes in the initial conditions. And that, strange attractors can be found that seek to take the system to previous states or future desired states with less complexity.

Decision-makers in Social Complex Systems must not be fooled by any person or company who peddle fads and quick fixes for complex systems. The problem situations they have to face as owners, or heads of those complex systems are too complicated and diverse to be handled by anything other than the systemic and creative form of intervention that critical Systems Thinking and Critical Systems Practice seek to guide and structure.

Critical Systems thinking tools and Critical Systems Practice advocates and enables the maximum creativity when the problem situation is being analyzed.

Critical Systems Thinking and Critical Systems Practice helps decision-makers to evaluate the usefulness to them in their particular situation, of different decision-making solutions and particularly different systems methodologies and methods, in addition, to use them in combination if it is needed.

Critical Systems Thinking and Critical Systems Practice points to the need to evaluate decision-making in the long term by using different types of measures and tools efficiently and effectively, the exploration of purpose and growth of mutual understanding, viability, anti-fragility, empowerment, emancipation, sustainability, and possible unintended consequences.

Because their methods are based on Action Research (AR) philosophy, Critical Systems Thinking and Critical Systems Practice provide a learning system that decision makers can tap into, in order to reflect on what has been achieved and how to further improve their own practice.

I hope in the future Critical System thinking will be taught more actively not just in undergraduate programs but also to graduates, professional students, and executives who can benefit from them; in different terms of time either on short courses or for many years. People who are exposed to System Thinking Study cases gain incredible insight from viewing those case studies. Each perspective allows them to see different things and to supply different explanations for the same phenomenon very fast. The value of Critical Systems thinking become evident. Becoming familiar with CST methodologies and methods allows its users to have an impact on the "*Worldviews or Welstangshaugns*" of the decision-makers and so the value of soft systems approaches becomes evident. As problems situations change a different choice of methodologies may be necessary to show the flexibility and versatility of CST.

I hope this book has the power to motivate readers to change the world they are a part of, by using the power of complexity critical systems thinking tools and Complexity Theory to support the development of the communities they are a part of.

I also thank you for taking the time to read my ideas and learn about my work. Hopefully these ideas and these jobs will inspire you to develop new ideas to develop interesting jobs that change the lives of many people.

As we reach the end of this book, our journey through critical systems thinking and complexity theory has been nothing short of transformative. We have explored the intricate web of connections that shape our communities, the profound interplay of systems, and the beauty of embracing complexity in our pursuit of positive change. Throughout these pages, we have not only delved into theoretical concepts but also equipped you with practical tools to navigate the complexities of community development and peace-building. By combining clear explanations, engaging content, and a focus on tangible outcomes, our goal has been to empower you, our readers, to become agents of transformation in your communities. Now, armed with critical systems thinking and complexity theory, you possess a powerful lens through which to see the world, allowing you to decipher the hidden patterns, feedback loops, and leverage points within the systems around you. By understanding the interconnectedness of issues, you can address root causes rather than merely treating symptoms, leading to sustainable and lasting impact. But theory alone is not enough. Our ultimate aspiration is to inspire action, driving positive change at every level. We urge you to take what you have learned here and apply it in your community development and peace-building endeavors. Remember, it is the collective effort of committed individuals like you that can lead to profound shifts and long-lasting improvements. As you embark on your journey, always remember to embrace diversity, encourage collaboration, and involve the voices of those most affected by the challenges your community faces. Seek partnerships with like-minded individuals and organizations, for it is through collective wisdom and shared experiences that we can create a ripple effect of positive change. Keep learning, keep adapting, and keep challenging yourself to think critically and creatively. Complexity may present obstacles, but it also harbors opportunities for innovation and growth. Embrace uncertainty, for it is often the catalyst for transformation. Together, let us foster a deeper understanding of the interconnectedness and complexity of community development and peace-building endeavors. Let us work side by side to co-create solutions that address the unique needs and aspirations of each community, ensuring that no one is left behind. Our journey does not end here. It is an ongoing quest for knowledge, understanding, and compassion. As you step forward, know that you have the power to shape the world around you. Embrace this power responsibly, always guided by the principles of empathy, justice, and sustainability. Thank you for embarking on this journey with us. May your actions be a beacon of hope and an inspiration for others. As you make a difference in your community, remember that you are part of a larger movement towards a brighter, more connected, and peaceful world.

The end…

References

Ackoff, R. (1993). Idealised Design: Creative Corporate Visioning. Omega: The InternationalJournal of Management Science, 21, 401–410.

Braun, W. (2002). The System Archetypes. In: Braun, W., The Systems Modeling Workbook.

Capra F Sempau D (1998) The Web of Life.

Child, J. (1972): Organization structure, environment and performance. The role of strategic choice. Sociology, 6, 1–22.

Churchman, C.W. (1970). Operations research as a profession. Management Science, 17, B37-53.

Churchman (1974) Perspectives of the systems approach

Churchman CW, Ackoff RL, Arnoff EL. 1957. Introduction to Operations Research. John Wiley: New York.

Forrester, J.W. (1961). Industrial Dynamics. MIT Press, Cambridge MA.

González, Mª Pilar (1987): Orientación sociométrica. En Mª. Pilar GONZÁLEZ y Esteve VENDRELL (Dirs.): El grupo de experiencia como instrumento de formación. Barcelona: PPU.

Jackson, M. (2019) Critical Systems Thinking and the Management of Complexity. Responsible leadership for a complex world.

Jackson, M.C and Keys, P (1984) towards a system of systems methodologies. Journal of the operational research society 35:473-486.

Johnson N (2007) Simply complexity: A clear guide to complexity theory

Meadows DH (1972) Limits of Growth : Original edition MIT Press, Cambridge MA.

Midgley (1992). Pluralism and the legitimation of systems science. Systems Practice, 147–172. & the sacred and profane in critical systems thinking. Systems Practice, 55–16.

Midgley, G. (1995). What is this thing called Critical Systems Thinking? In K. Ellis, B. Mears-Young, & G. Ragsdell, Critical Issues in Systems Theory and Practice. New York: Plenum

Midgley, G. (1997) Mixing methods:developing systemic interventions.In:Multimethodology (Mingers & Gill) p291–332 Chichester Wiley

Midgley, G., Munlo, I., & Brown, M. (1998). The Theory and Practice of Boundary Critique: Developing Housing Services for Older People. Journal of the Operational Research Society.

Midgley, G. & Pinzón, L. A. (2011). Boundary critique and its implications for conflict prevention. Journal of the Operational Research Society, 62, 1543–1554.

Midgley, G., & Pinzón, L. A. (2013). Systemic Mediation: Moral reasoning and boundaries of concern. Systems Research and Behavioral Science, 30, 607-632.

Navarro, J. (2001). "Las organizaciones como sistemas abiertos alejados del equilibrio". Tesis de doctorado no publicada. Universidad de Barcelona: España.

C. A. Forero Navarro, *Untangling Complexity—Peace Building Engineering*, Synthesis Lectures on Engineers, Technology, & Society 29, https://doi.org/10.1007/978-3-031-37038-0

Navarro, CA (2022) Untangling Complexity Complexity Funnel Methodology & Social Transformation Workshops Universidad de los Andes Bogotá DC Colombia
Peiró, J.M. (1990): Organizaciones. Nuevas Perspectivas Psicosociológicas. Barcelona: PPU.
Senge PM (2005) The fifth Discipline in practice
Stacey, RD (1992) Managing the unknowable: Strategic boundaries between order and chaos in organizations
Sterman J. 2000. Business Dynamics. Irwin-McGrawHill: New York.
Ulrich, W. (1987). Critical heuristics of social systems design. En W. Ulrich, European Journal of Operational Reasearch (págs. 276-283). North-Holland: Elsevier Science Publishers.
Varela, Francisco J.; & Maturana, Humberto R. (1973). De Máquinas y Seres Vivos: Una teoría sobre la organización biológica. Santiago de Chile: Editorial Universitaria. Hay edición revisada de 1995.
Weik, Karl E. (1969): The social psychology of organizing.

Uncited References

Ackoff, Russell L Difference that make a difference (2010).
Ackoff, RL., Magidson, J., and Addison, H.J. (2006) Idealized Design :Creating an Organization's Future. Philadelphia, PA: Wharton School.
Beer, S. (1966). Decision and Control: The Meaning of Operational Research and Management Cybernetics. Chichester: Wiley.
Burns, T. y Stalker, G. M. (1961): The management of innovation. Londod: Tavistock Publications.
Checkland P.B & Holwell, S (1998–2004) Clasic OR and Soft OR an asymmetric complementarity.In:Systems Modelling :Theory and Practice (ed M.pidd) 44–60.
Hoverstadt, P (2010) The Viable System Model in Systems Approaches to Managing Change eds Reynolds, M. and Holwell, S
Jackson, M.C. (2006). Beyond problem structuring methods: reinventing the future of OR/MS.
Perez Rios J, Design and Diagnosis for Sustainable Organizations, https://doi.org/10.1007/978-3-642-22318-1_3, Springer-Verlag Berlin Heidelberg (2012).

The manufacturer's authorised representative in the EU is Springer
Nature Customer Service Centre GmbH, Europaplatz 3, 69115 Heidelberg,
Germany. If you have any concerns regarding our products, please
contact ProductSafety@springernature.com

Printed and bound by CPI Group (UK) Ltd, Croydon, CR0 4YY
01/10/2026
02225904-0020